O-1 Bird Dog in action

By Al Adcock

Color By Don Greer

Illustrated By Joe Sewell

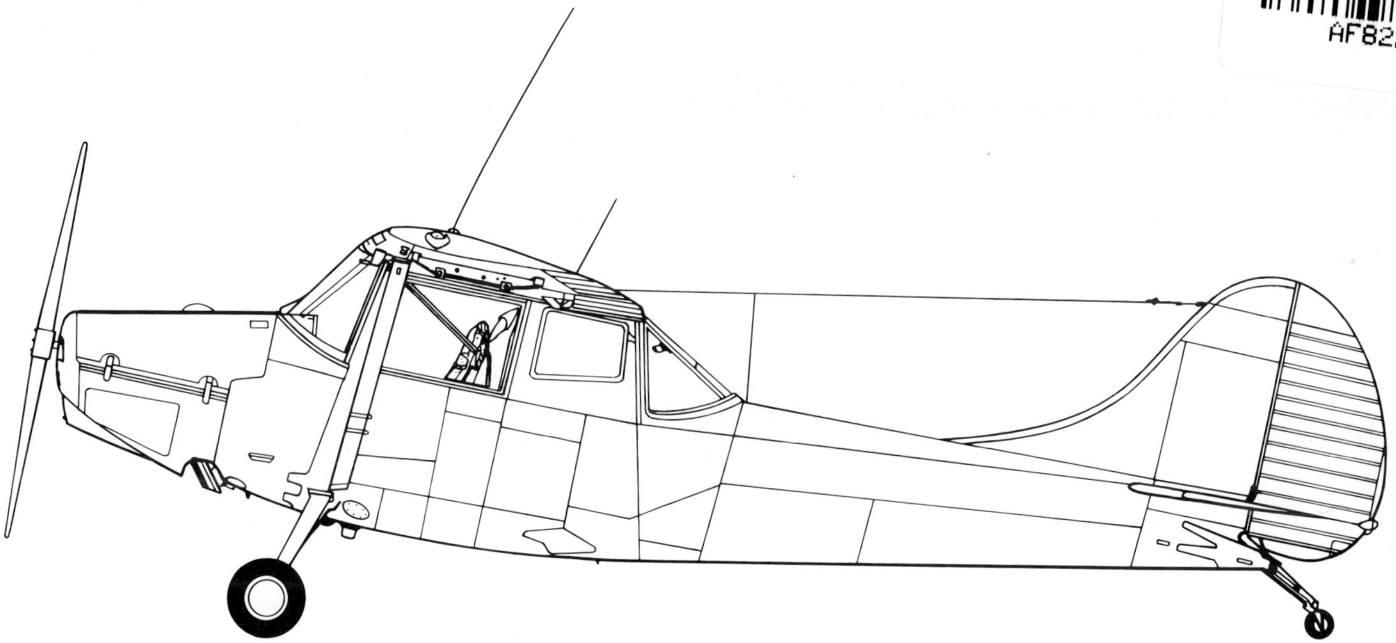

Aircraft Number 87

squadron/signal publications, inc.

A TO-1D Bird Dog of the 73rd Aviation Company pulls up after marking a target for tactical fighters near the village of Bon Sar Pa, South Vietnam during March of 1964.

Cessna

ISBN 0-89747-206-3

If you have any photographs of the aircraft, armor, soldiers or ships of any nation, particularly wartime snapshots, why not share them with us and help make Squadron/Signal's books all the more interesting and complete in the future. Any photograph sent to us will be copied and the original returned. The donor will be fully credited for any photos used. Please send them to:

Squadron/Signal Publications, Inc.
1115 Crowley Drive.
Carrollton, TX 75011-5010.

DEDICATION

This book is dedicated to the people who gave me the encouragement to achieve my goal of telling the O-1 Bird Dog story. My wife, Carol, who labored many hours typing and editing my manuscripts. My two sons, Albert and Chris, for their support and especially Tim Edwards of the US Army Aviation Museum, Fort Rucker, Alabama for his invaluable help.

This book is further dedicated to the aviators and crew chiefs that were associated with the O-1, especially those who made the ultimate sacrifice; and to Bob Pickett and Charles Blackman, both of whom passed on since assisting me with this project.

Acknowledgements

My very special thanks to the following people and organizations for their help in obtaining photographs and information on the O-1 Bird Dog.

Chris Adcock
Garth Adcock
Arnold Bell
G. Bennett
N. Berlinghoff
B.E. Best
Charles Blackman
Boeing Archives
Rex Boggs
Canadian Forces
Earl Caudell
Cessna
C. Corn
Howard Corns
Albert Devito
A. J. Dorn
John Elliott
Bill Grant
Jim Groves
L. Hickey
Jack Jaubert
Bill Johnson
G. Kamp
John S. Kark
Jim Kitchens
F. LePage
A. J. Lavalle
T. Love
C. Maranta
David Menard
Daniel Moraga
Hugh Morgan
Mike Mullins
Nat'l Aviation & Space Museum
Peter Palo
Tim Parker
Marilyn Phipps
Robert J. Pickett
A. Robinson
Robert E. Standland
G. Stanton
M. Swartzberg
Norm Taylor
Teledyne CAE
US Air Force
US Army
US Army Aviation Museum
US Army Military History Institute
US Marine Corps
US Naval Aviation Museum
Bob Woodford
Clifford Yarbrough

An Army O-1D patrols over the South Vietnamese coast during 1966. The aircraft has Orange identification panels on the upper wing and White painted flaps as a visual aid for strike aircraft. (F. LePage via R. Pickett)

U.S. ARMY
72901
ARMY

INTRODUCTION

During the late 1940s the U.S. Army began a search for a light observation aircraft to replace their aging fleet of Piper L-4 and Stinson L-5 aircraft. Both of of these earlier aircraft were fabric covered and the Army was now seeking to replace them with an all metal design.

Cessna, founded by Clyde V. Cessna in 1911, was primarily a manufacturer of civil aircraft for the general aviation market and had never seriously pursued a military contract. During the war Cessna had built several aircraft for the military, however, they had been adaptations of civil aircraft such as the AT-17 advanced trainer, or had been built under sub-contract such as the Waco CG-4A glider. In August of 1949 the Army, in cooperation with the Air Force, issued a joint specification for a new light observation aircraft. Cessna decided to respond to this specification and began researching just what type of aircraft would best meet the Army/Air Force requirement.

The joint specification called for an all metal light aircraft that could operate from short unprepared fields and perform a variety of missions. The roles envisioned by the Army and Air Force included; observation and visual reconnaissance, rescue, photographic reconnaissance, re-supply of ground troops, aerial telephone wire laying, artillery fire control, transportation of personnel over front line areas, message pickup and delivery, pilot training, casualty evacuation, Forward Air Control of fighter-bombers, and observer training. Additionally, the specification stated that the aircraft should be adaptable for both ski and float landing gear.

The Army and Air Force required a 'finished article' prototype be produced for a fly-off competition that would be held among all competing designs presented for evaluation. The projected date for the evaluation fly-off was set for March of 1950.

Cessna designers, Derby Frye and Jerry Gerteis, began work on the Cessna contender by visiting various Army posts and the Pentagon to find out just what the Army really needed in an observation aircraft. From these discussions it became obvious that the Army wanted an aircraft that was able to carry a reasonable amount of radio equipment, would be relatively simple to maintain, and could fly in and out of short unprepared airfields near the front lines.

During 1949, the Cessna Engineering Department consisted of eighteen people engaged in various civilian aircraft projects. The engineers knew that to meet the deadline set by the Army for the competition fly-off it would be necessary to use components from existing Cessna production aircraft. By using the wings from the Model 170 and the tail assembly of the Model 195 the engineering staff was able to concentrate on the fuselage, landing gear, and engine installation which were of all new design. By the use of these proven components, Cessna was able to begin construction of the prototype on 8 September 1949. Work on the Model 305 prototype progressed smoothly and was completed ninety days later, after some 2,500 man hours of design work by the engineering staff. The prototype rolled out of the experimental hanger in Wichita on 8 December, ready to begin factory flight tests.

The Model 305 was a two seat, all metal, high wing monoplane fitted with a conventional landing gear and an all around vision cockpit. The prototype was constructed entirely of aluminum, making it both rugged and light weight. The fuselage assembly was of semimonocoque rivetted metal construction utilizing aluminum alloy bulkheads, stringers and an aluminum alclad skin covering. The thirty-six foot wing was a conventional stressed skin, strut braced, two spar, semi-cantilever construction assembly covered by aluminum alclad skinning.

All control surfaces including the ailerons, rudder, flaps, and elevators were also metal framing with alclad aluminum skin. This construction method allowed the Model 305 to be operated from rough unprepared surfaces and also made the 305 easy to repair in the field under any weather conditions. The prototype was stressed for 4.4 Gs and a maximum dive speed of 198 mph. The inboard wing sections each held a twenty-one gallon fuel tank fitted with direct reading fuel gauges in the wing roots, which were readily visible to the pilot or observer in the cockpit.

The Piper L-4 was the first Army Air Corps Forward Air Control (FAC) aircraft. The L-4 was also used in the liaison, trainer, and observation roles in two wars, World War II and Korea. Piper built almost 5,000 L-4s and it was the most widely used liaison aircraft in US service. (U.S. Army Aviation Museum)

The prototype was powered by a 213 hp Continental E-190 (also known as the O-470) six cylinder horizontally-opposed engine driving a metal two blade, fixed pitch, McCauley propeller. The excellent power to weight ratio of the Continental engine allowed the prototype to have a high cruising speed, yet retain sufficient power reserves for emergencies. The twenty-four volt DC electrical system consisted of a fifty amp generator driven by the engine and a twenty-four volt battery that was located between the rudder pedals in the cockpit. An external power source plug-in jack was mounted on the starboard forward fuselage side, easily accessible to the ground crew.

Responding to a joint Army/Air Force request for a light liaison aircraft to replace the Piper L-4 and the Stinson L-5, Cessna produced the prototype Model 305 (N41694). The prototype was completed within ninety days and then flown to Wright Field in Ohio to participate in a competitive 'fly-off' against other designs. (Cessna)

The Model 305 prototype (N41694) in December of 1949 after being modified with steps on the forward fuselage just above the landing gear leg and an external power source receptacle on the forward fuselage behind the cowling. The prototype used the wings from the Model 170 and a Model 195 tail assembly. (R. Pickett)

The landing gear wheels could be fitted with standard 7:00 X 6 tires or optional oversize tires (8:00 X 6) for use on unimproved grass or dirt fields. The tail wheel size remained the same regardless of the tires used on the main gear.

The pilot and observer sat in tandem with the radio gear installed behind the observer's seat. To allow for maximum visibility the cockpit was fitted with large plexiglass panels completely circling the cockpit. The windshied and rear window were single piece curved plexiglass panels. The cockpit door on the right side of the fuselage contained a large single piece window which was hinged to open upward and outward and could be locked open in flight by means of a retaining stud on the undersurface of the wing. The rear side windows opened inward and upward and could be strapped to the cockpit ceiling. Additionally there were six small windows set into the center of the cabin roof. The pilot and observer were provided with dual sets of rudder pedals and control sticks. The observer's stick could be removed when not in use and stored in a special storage rack within the cockpit.

The prototype weighed 1,400 pounds, some two hundred pounds heavier than the Army specification which called for an aircraft of 1,200 pounds. Performance, however, was well within the specification making the weight problem unimportant.

After completion, the prototype was given a factory flight test before being ferried to Wright Field by Hank Waring, Cessna's Chief of the Engineering Flight Test and Aerodynamic Group. The Model 305 prototype was entered in the Army fly-off competition against competing designs from Piper, Temco, and Taylorcraft, who had been supplying the Army and Air Force with light aircraft for years. The fly-off competition was held over a six week period, during which Waring quickly found that the hardest part of the Army specification was the requirement for landing over a 50 foot obstacle in less than 600 feet. The requirement was finally met by landing with sixty degree flaps, a high sink rate, and a touchdown on the main wheels.

On 29 May 1950, Cessna was verbally notified that they would be awarded an Army production contract for 418 aircraft under the designation L-19. The Army desired a delivery of production units in September, however, and that was out of the question since Cessna had not received a type approval certificate from the Civil Aeronautics Administration (Federal Aviation Administration today).

On 25 June 1950, the North Korean army crossed the 38th Parallel invading the Republic of South Korea beginning the Korean War. With the United States in another shooting war, new demands were placed on Cessna by the Army to increase production. These demands would not only be met — they would be exceeded.

The Cessna L-19 Bird Dog would go on to be built in five different models and be known under a number of different designations. Initially Navy/Marine Corps aircraft were designated as OE-1s. Later, when the Department of Defense standardized aircraft designations, the L-19/OE-1 designations were dropped and all Bird Dogs became O-1s regardless of service. The L-19/O-1 was an aircraft that produced a number of firsts. It was the first all metal high wing aircraft to service with Army Aviation; the first turboprop light aircraft to fly over 35,000 feet; and the first of what would become many military aircraft built for the US Armed Forces by Cessna.

Eventually the Army would accept a total of 3,105 L-19s and Cessna would also produce a number of other Bird Dogs for export with the L-19/O-1 serving in a number of foreign air forces. The Bird Dog proved extremely popular with the Japanese and to meet the needs of the expanding Japanese Self Defense Forces, Fuji Heavy Industries contracted with Cessna to assemble L-19s under license in Japan.

It is obvious that the US Army made the right decision when it chose the L-19 as their new liaison/observation aircraft. The name 'Bird Dog' was given the aircraft by GEN Mark Clark from a list of winning entries submitted by Cessna employees. The General originally wanted to name the L-19, 'Skyhawk', however, the Douglas A-4 had already laid claim to the name. So the next most popular name, 'Bird Dog' was declared the winner. The name aptly fits the L-19 and its mission of searching out and marking enemy targets, similar to the way a bird dog searches out and points at a bird. The L-19 is one of the most enduring aircraft of modern military history, and is still serving with a number of foreign countries and the US Civil Air Patrol.

Fitted with a pair of EDO floats, the 305 prototype demonstrates its versatility for Army evaluators. The 'fly-off' pitted Cessna against Piper, Taylorcraft, and Temco for the Army contract. The prototype could also be fitted with skis, tandem landing gear, and cross wind landing gear to allow it to operate from any type of surface. (R. Pickett)

Development

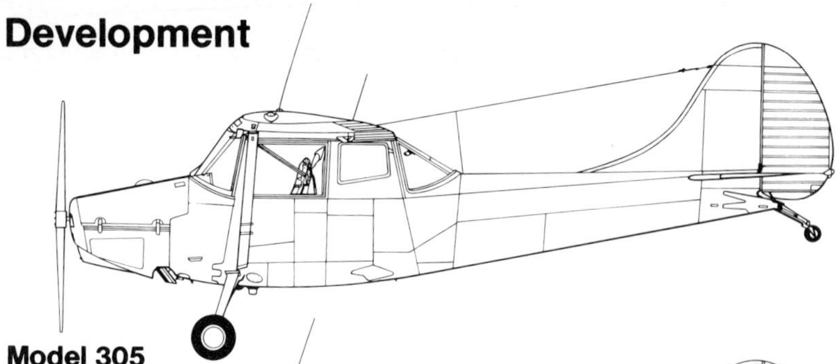

Model 305 Prototype

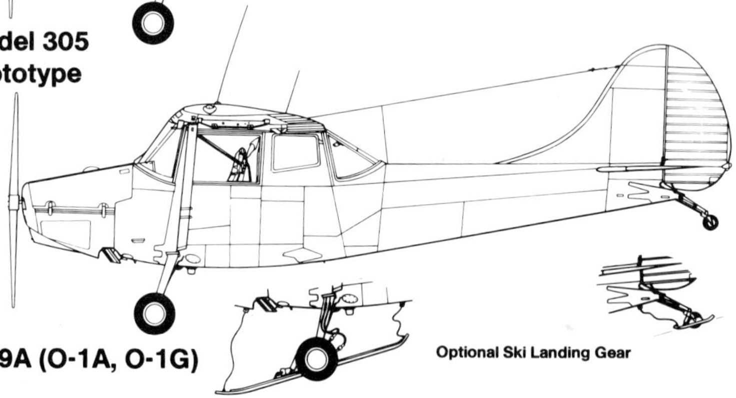

L-19A (O-1A, O-1G)

Optional Ski Landing Gear

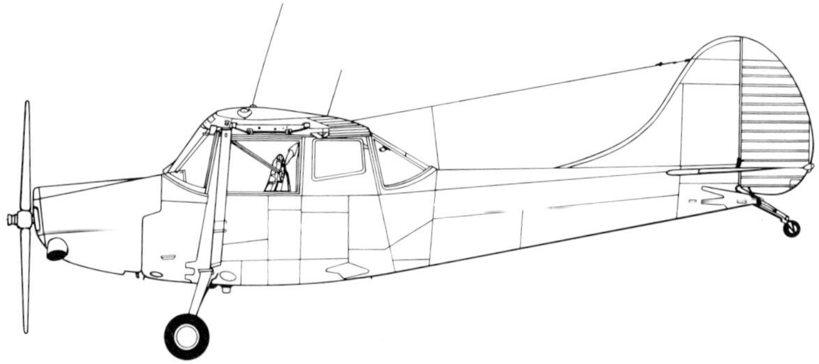

XL-19B Prototype

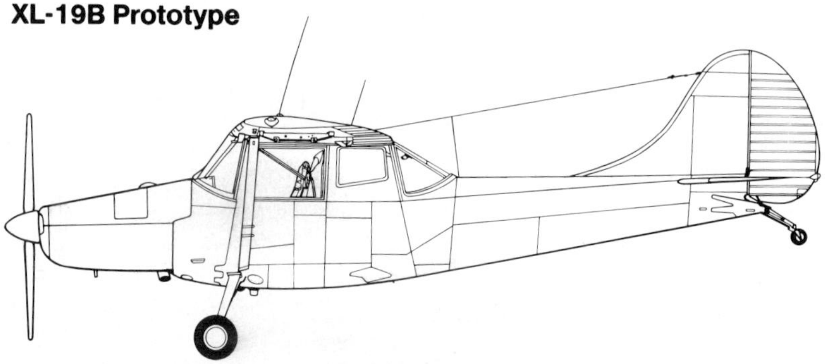

XL-19C Prototype

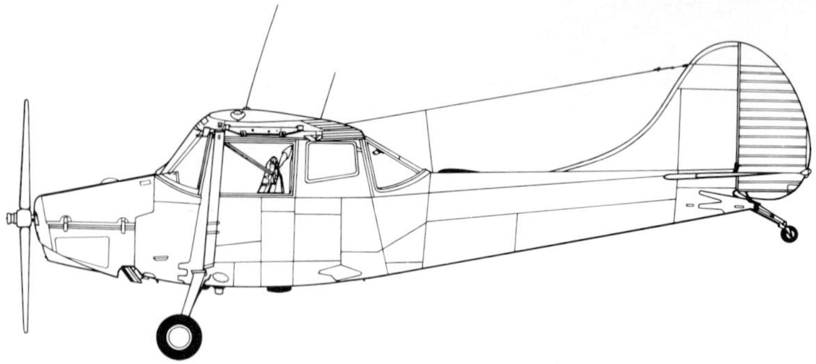

TL-19D (O-1D)

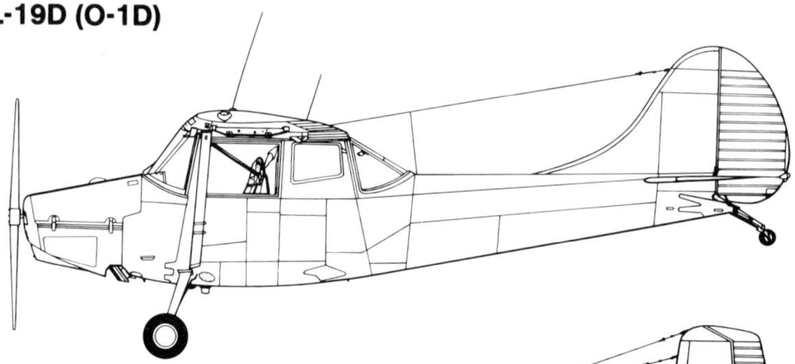

L-19E (O-1E)

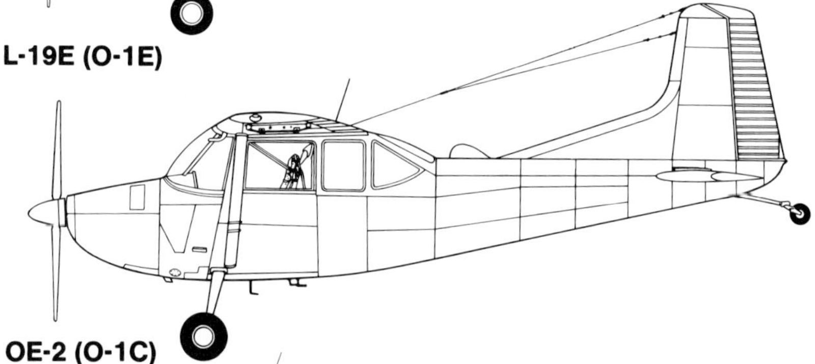

OE-2 (O-1C)

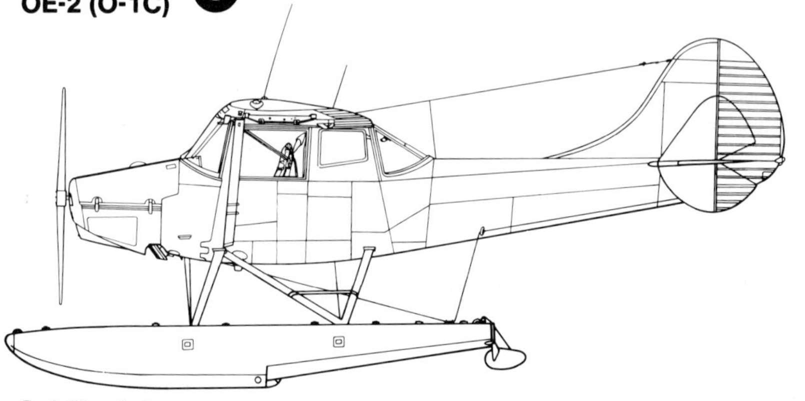

O-1 Floatplane

L-19A (O-1A)

Production L-19As were externally identical to the prototype with the exception of the installation of two underwing Mk 4A bomb racks under the outer wing panels just outboard the wing strut. These racks were capable of handling underwing loads up to 250 pounds per wing and were normally loaded with four 2.75 inch smoke rockets. These rockets were used to mark targets for high speed fighter-bombers.

Production L-19s also had provision for installation of either ski and float landing gear. When properly corrosion-proofed for seaplane operations, a pair of Edo floats could be installed in place of the normal landing gear. The float installation could be accomplished in the field, by removing the main and tail wheels and attaching the float supports to the main gear attachment points. For use on ice and snow the L-19A could also be equipped with Federal Aircraft Works skis, which mounted directly to the main landing gear legs. The skis have cutouts for the wheels making it unnecessary to remove the main gear in order to mount the skis. With these provisions, the L-19A was capable of operating from any type of surface, anywhere in the world.

The first L-19 (serial 50-1327) was delivered to the Army on 20 December 1950 and was used initially for pilot training before seeing combat in Korea. The pressures of the Korean War led the Army to quickly deploy L-19s to Korea where the Bird Dog made its combat debut on 16 February 1951. Army pilots readily appreciated the maneuverability, speed, and ruggedness of the Bird Dog and it became an instant favorite for reconnaissance, artillery spotting, and transporting various Very Important Persons (VIPs) on inspection trips over the front.

During one such trip, the first production L-19A was used to transport the President-elect, Dwight D. Eisenhower on an inspection tour during 1951. This aircraft had a particularly unique history; it was used as the base model for writing the L-19 Pilot's handbook and during June of 1952 was the first L-19A turned over to the Republic of Korea Army for training and operational missions. Later this same L-19 served as the personal liaison aircraft for General Choi Yong Dok, Chief of Staff of the South Korean Army. In June of 1954 the aircraft was returned to American control serving with the Air Section of the US Army North Corps, 1st Cavalry Division. Finally during 1959 the aircraft was retired and passed to the US Army Aviation Museum at Fort Rucker, Alabama where it still proudly serves as a tribute to the pilots who flew the L-19. Other L-19s served as the personal observation aircraft for a number of high ranking Army officers including; Generals Van Fleet, Ridgeway, and Taylor.

The L-19A was the most widely produced Bird Dog variant with a total production run of 2,578 aircraft. The majority of Bird Dog production went to the US Army, however, a number of other countries, seeking replacements for their aging liaison aircraft, found the L-19A well suited to their needs. Chile was the first foreign country to add the L-19A to its inventory taking delivery of six L-19As under the Military Assistance Program.

France received their first L-19s during the 1950s taking delivery of some ninety aircraft for duty with the French Army. Canada purchased sixteen L-19As for the Royal Canadian Air Force, receiving their aircraft near the end of the L-19A production run.

The U.S. Marines took delivery of thirty-five L-19As borrowed from the Army under the designation OE-1. These aircraft were assigned to Marine Observation Squadron Six (VMO-6) for duty in Korea.

During 1953, the Army had a number of L-19As in Japan and Korea. Rather than return these aircraft to the US, they were given to the Japanese Air Self Defense Force where a number continue to serve today.

The first production L-19A (50-1327) flies over the open fields around Wichita, Kansas in December of 1950. The aircraft was delivered in natural metal with Black anti-glare panels and markings. The last three digits of the serial number preceded by an 'A' for Army was carried on the upper right and lower left wing. (U.S. Army Aviation Museum)

During the mid-1950s the US Military Assistance Command in Vietnam (MACV), with the assistance of the French, supplied the Republic of South Vietnam with two squadrons (twenty aircraft) of L-19As for their fledgling air force. Initially these aircraft were used for pilot training, liaison, and personal transports for high ranking Vietnamese officers. By 1969 the Vietnamese Air Force (VNAF) had increased its inventory of L-19As to over sixty aircraft with the majority committed to the Forward Air Control (FAC) role.

Cessna held a company wide contest to select a name for the L-19. This L-19 on display outside the Cessna Aircraft Factory in Wichita during August of 1951 was used to promote the contest. The name 'Bird Dog' was chosen from the winning entries by GEN Mark W. Clark, Chief of Army Field Forces. (R. Pickett)

L-19 Boundary Layer Control Test Bed

During 1955, the Army provided an L-19A (serial 51-11972) to the Aero Physics Lab at Mississippi State University (MSU) for service as a test bed for flight tests involving investigation of the Boundary Layer Control (BLC) theory. As part of the experiment, the L-19 was modified with a tricycle landing gear installed by the engineers at MSU. The nose gear was taken from an early Piper Tri-Pacer and the main gear was the standard L-19 gear rotated 180 degrees. Mel Swartzherg, the chief test pilot at MSU, stated that he had very little faith in the nose gear and primarily landed the aircraft on the main gear, keeping the nose gear off the ground until the aircraft was almost stopped. He reported that the modified L-19A could be flown at airspeeds as low as 29 mph and the aircraft was almost impossible to stall. Using Boundary Layer Control the aircraft could be landed and brought to a stop within 150 feet.

George Bennett, who was involved in the project, explained how the Boundary Layer Control worked. High pressure air, blown over the flying surfaces, greatly increases lift enabling the aircraft to remain flying at greatly reduced airspeeds. The system also allows the aircraft to land and take off in far shorter distances than normally possible. MSU flew 400 hours testing the L-19A during the seventeen year project with the Bird Dog undergoing a number of modifications.

Since the aircraft was going to be routinely operated at reduced landing speeds, it became necessary to increase the area of the vertical stabilizer to enhance rudder control. Other modifications included reshaping the rear windshield, addition of boundary layer fences on the horizontal stabilizer, and installation of a Continental 0-470-15 engine and McCauley constant speed propeller. The engine/propeller combination gave the pilot a wider range of throttle and prop settings enhancing control and fuel consumption. The experiments were concluded in 1972, with the aircraft placed in storage until 1986, when it was sold back to the Army Aviation Museum for $1.00. Today the aircraft is on display at Fort Rucker, Alabama.

All of the production L-19A models were equipped with hydraulically operated flaps. Reports from combat pilots revealed that a single bullet hit could render the flaps inoperative, severely restricting maneuverability and landing performance. To correct this problem the Army modified one L-19A with electrically activated flaps. Tests with the new flaps prove successful and it was decided to modify the entire L-19A fleet with electric flaps. To facilitate the conversions, Cessna produced modification kits that could be easily installed by Army maintenance crews in the field. The electrically operated flaps proved to be more serviceable by maintenance crews, better able to resist battle damage, and were easier to operate.

The build-up of U.S. forces in South Vietnam during the early and mid 1960s led to increased demands by field commanders for liaison and reconnaissance aircraft. To meet these demands a number of L-19As that were in storage were refurbished and put back into active service as O-1As, redesignated under the standardized designation system that had come into use during 1962. A number of other L-19A/O-1As, which had been modified as two seat instrument trainers with dual controls and instrument panels under the designation TL-19A/TO-1A, were again modified by removing the rear instrument panel and strengthening the outboard wings to handle installation of an additional Mk 4A bomb rack. These modified aircraft were re-designated as O-1Gs under the standardized designation system. L-19A/O-1A/O-1Gs served throughout the Vietnam War, alongside later models of the Bird Dog, being flown by all US services and the South Vietnamese Air Force (VNAF).

The assembly floor at Cessna during 1954 shows L-19s in various stages of construction and a Cessna XH-41 Seneca helicopter against the far wall. After final assembly, the Bird Dogs were taken to the paint shop where the paint scheme and markings were applied. (R. Pickett)

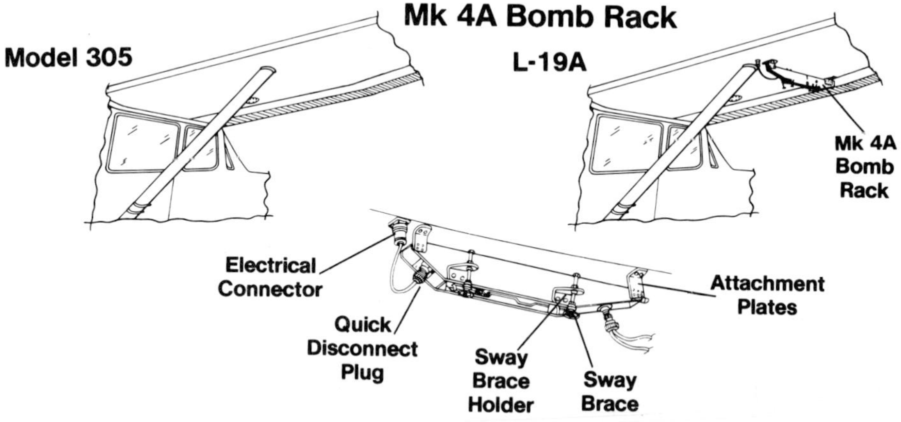

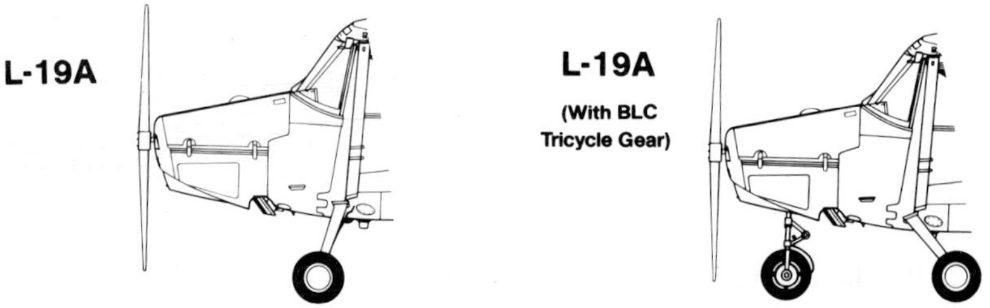

In a demonstration of its short field landing capabilities this L-19A clears a fifty foot obstacle before landing. The L-19A normally flew low and slow and could land within 600 feet. This L-19 is finished in overall Olive Drab. (U.S. Army Aviation Museum)

The port side opening cowl panel and O-470 Continental engine installation on an overall Olive Drab L-19A. The serial and servicing instructions are in Yellow. The 'G' in the serial indicates that this aircraft was procured by the USAF for the Army. (Boeing)

Army National Guard units around the country were re-equipped with L-19As to replace their aging L-4s and L-5s during 1952. This L-19A (51-4876) would later be modified during 1968 by removing all trainer equipment being re-designated an O-1G. (USAF via H. Morgan)

A new production L-19A (50-1451) prepares for takeoff from the Cessna plant for a delivery flight to an 8th Army unit. The aircraft is in overall natural metal with Black lettering and anti-glare panels on the nose and fuselage spine. (David Menard)

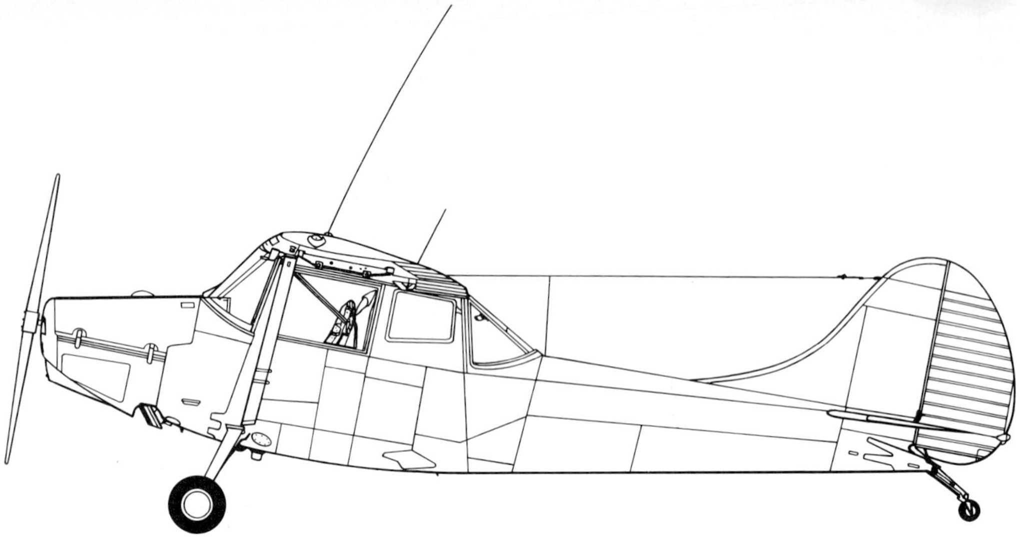

Specifications

Cessna L-19A (O-1A) Bird Dog

Wingspan	36 feet
Length	25 feet 9 inches
Height	7 feet 6 inches
Empty Weight	1,400 pounds
Maximum Weight	2,200 pounds
Powerplant	One 213 hp Continental O-470-11 six cylinder air cooled engine
Armament	Four 2.75 inch smoke rockets
Performance	
Maximum Speed	146 mph
Service ceiling	21,300 feet
Range	840 miles
Crew	Two

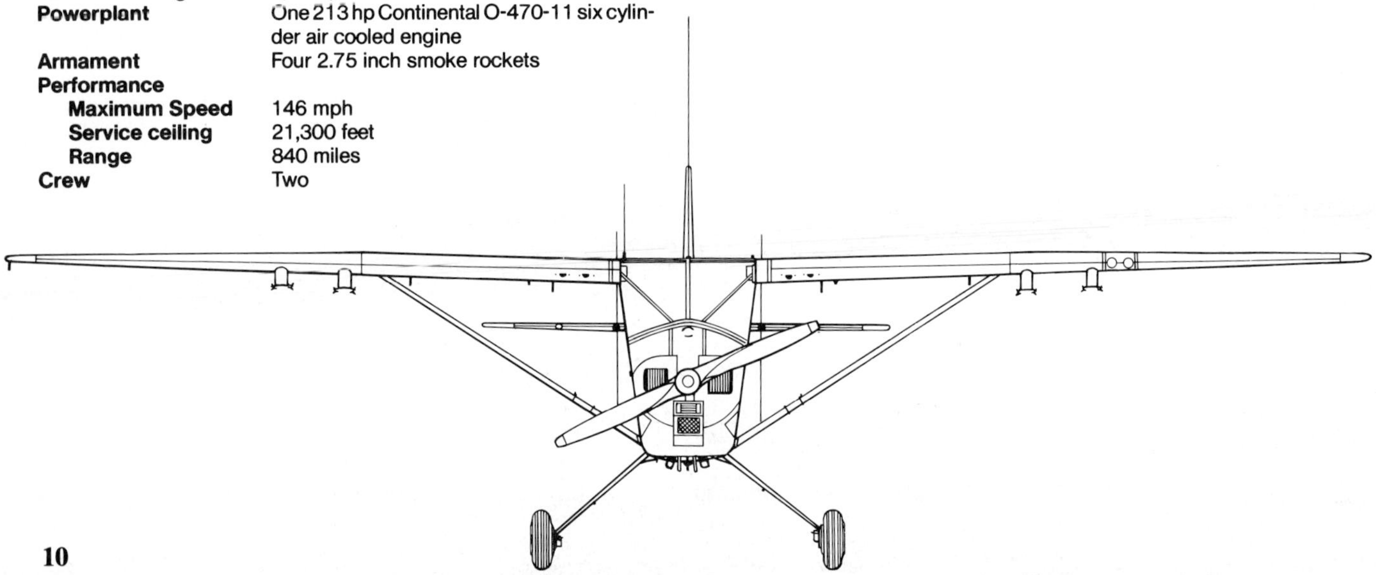

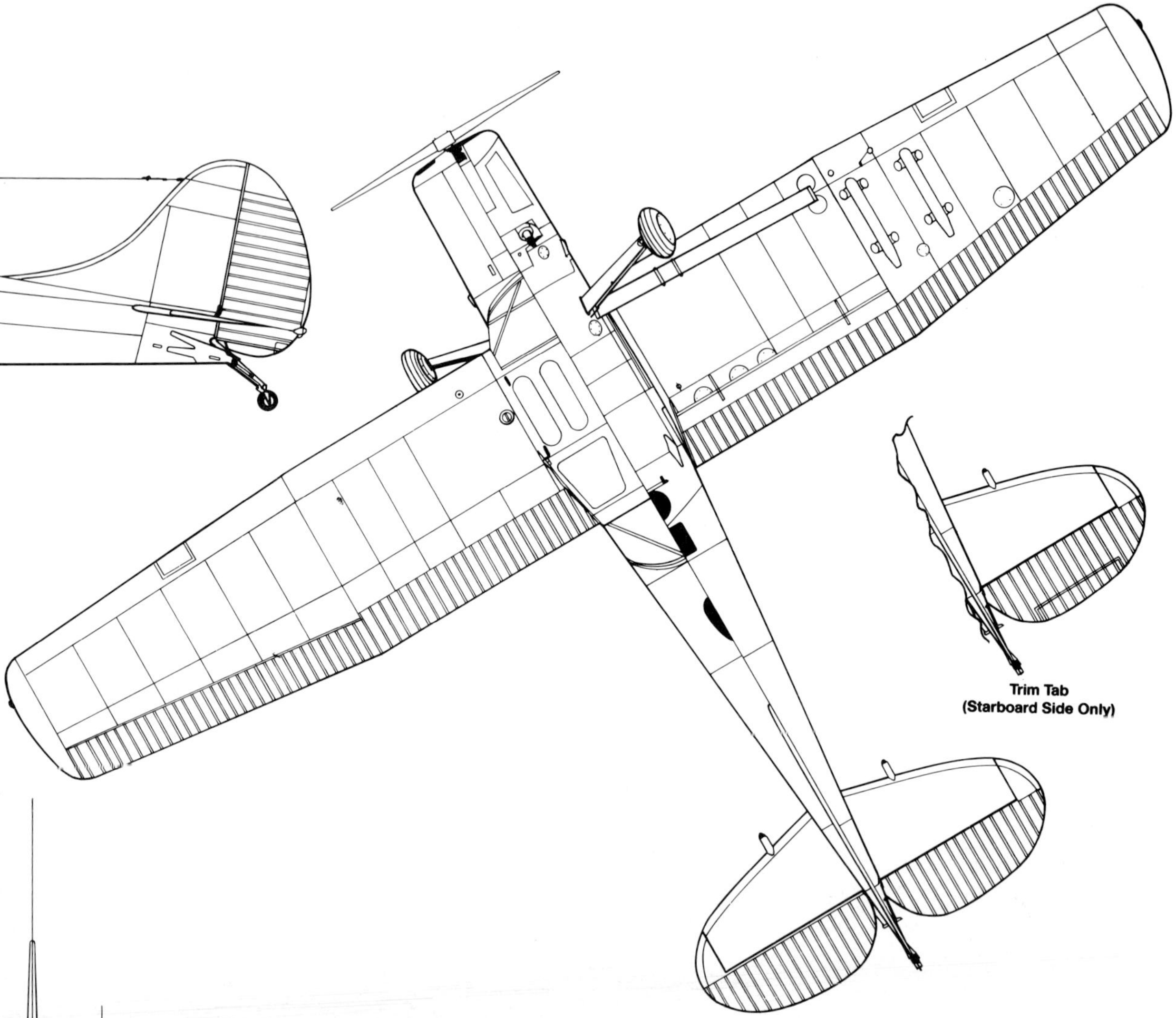

Trim Tab
(Starboard Side Only)

MAJ Wilfred Jaubert stands beside the L-19A (50-1327) that was used to fly President elect Dwight D. Eisenhower on an inspection tour over the Korean front lines during 1952. Before being sent to Korea the aircraft was painted overall Gloss Olive Drab with Yellow markings. The placard on the cowl is Red with five Silver stars, indicating a General of the Army was on board. (J. Jaubert)

After a landing accident that damaged the top of its rudder and fuselage this L-19A (51-4822) is airlifted by a US Army H-34 to a repair facility. L-19s were used in the basic training role to familiarize new pilots and accidents were common. (U.S. Army Aviation Museum)

This L-19A (51-4632) sits on a Korean airfield awaiting repairs to its damaged starboard wing. The L-19 proved to be an excellent observation aircraft with its high wing and 360 degree field of view, however, its slow speed made it vulnerable to ground fire. (US Army Aviation Museum)

Painted in a high visibility paint scheme, this L-19A (50-1635) was used in an area where maximum visibility of the aircraft was essential. The Red tail and wing panels are painted over natural metal, with all lettering in Black. The significance of the insignia under the observer's window is unknown, although the inscription reads *IMPRIMIS*, meaning 'among the first'. (D. Menard)

The Army provided this L-19A (51-11972) to the Aero Physics Lab at Mississippi State University as a test bed for a series of tests on the boundary layer control theory. The tricycle gear was installed by the Aero Physics Lab and consisted of a Piper Tri-Pacer nose gear and reverse installation of the L-19s main gear. (U.S. Army Aviation Museum)

The BLC L-19A (51-11972) was modified with an enlarged rudder, boundary layer fences on the horizontal stabilizer, a O-470-13 Continental engine, and a McCauley constant speed propeller. After testing was completed the aircraft was refitted with conventional landing gear. The BLC L-19 was painted overall trainer (international) Orange with the wing uppersurface in White and the tail and nose tip in Day-Glo Orange. All lettering and the anti-glare panel are Black. (Author)

Tandem landing gear was installed on a number of L-19s to allow operations from unprepared strips. The modified landing gear offered outstanding landing and take off qualities. This L-19A carrying a parachute cargo delivery bag under the wing was modified with dual gear during 1967 and redesignated as an O-1G. (D. Menard)

Tandem Main Landing Gear

Standard Landing Gear **Tandem Landing Gear**

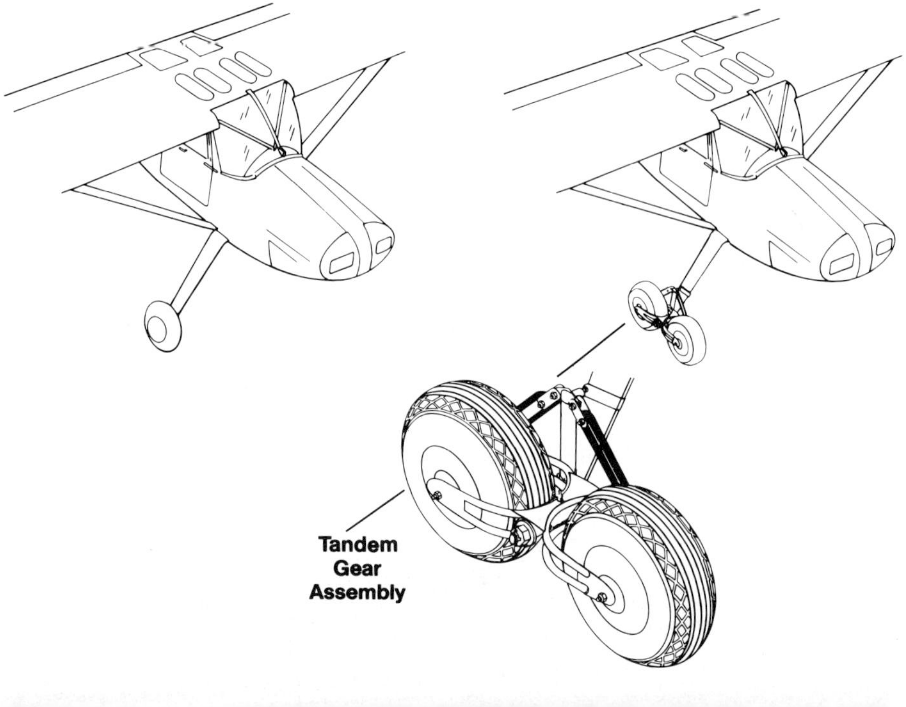

This L-19A (51-12304) at Lake Tholocco, Fort Rucker, Alabama, was used to demonstrate the universal landing gear (ULG) which consisted of two flat bottom skis attached to the main gear and a smaller ski on the tail wheel. The skis allowed the L-19 to land and taxi on water, however, the aircraft could not come to a complete stop until reaching land. (U.S. Army Aviation Museum)

Universal Landing Gear Installation

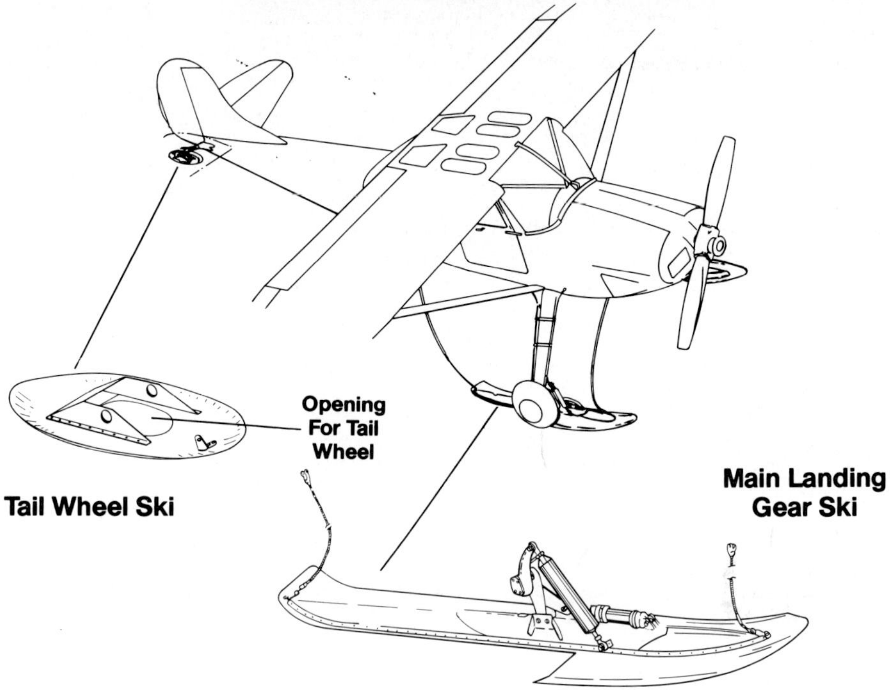

This L-19A (51-12711) assigned to Fort Rucker, Alabama is an instrument flight trainer with high visibility Orange panels painted over the Olive Drab camouflage to warn other aircraft that a student is in control. (US Army via C. Yarbrough)

This is the first production L-19A (51-4829) to be factory painted at Cessna in overall Olive Drab with Yellow lettering. An abbreviated serial number was painted on the upper starboard and lower port wing, however, this practice was discontinued during early 1953. (Cessna via C. Yarbrough)

This Army Reserve L-19A (51-12254) was one of a number of aircraft assigned to the Reserve and National Guard during the 1950s. Many of these aircraft were later converted to O-1G standards and reissued to regular Army and USAF units for duty in Vietnam. (D. Menard)

An L-19A (51-12227) undergoes radio tests at the Army Signal Corps Aviation Center, Monmouth County Airport, New Jersey in February of 1954. The ADF loop antenna on top of the cockpit roof was an early position that was later changed to the fuselage spine behind the observer's rear window. (National Aviation & Space Museum)

This L-19A (51-12394) on the ramp at St. Petersburg, Florida was used in the basic training role during 1958. The aircraft has had the unit insignia overpainted on the vertical stabilizer and is equipped with an FM antenna on the wing over the cockpit and an ADF loop antenna on the fuselage spine. (Author)

Antenna Configurations

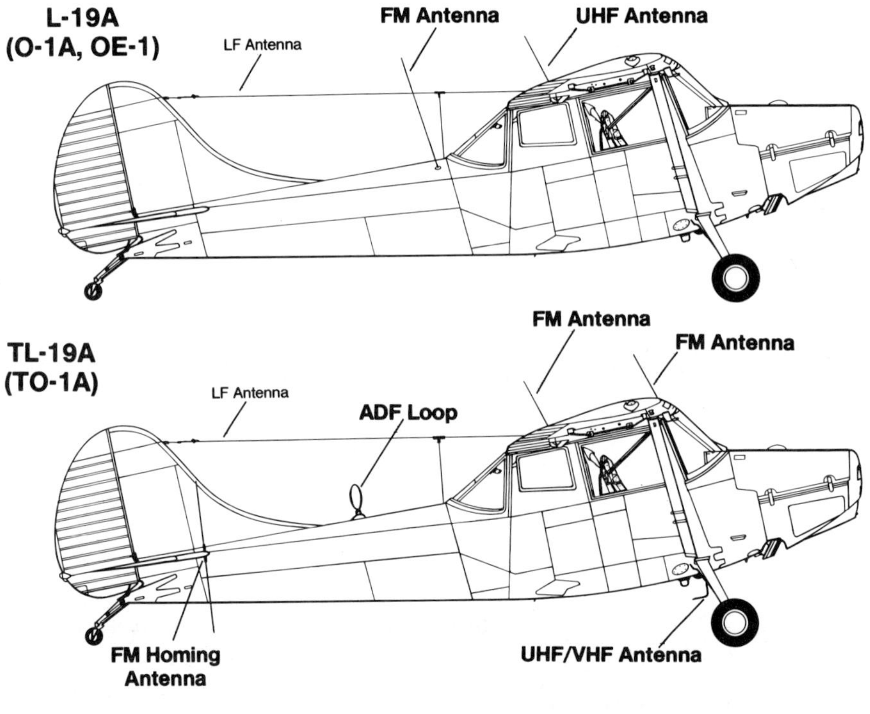

This early production L-19A (50-1525) is loaded with cargo bundles on early style underwing bomb racks for an aerial re-supply mission. The aircraft is finished in overall natural metal with Black anti-glare panel and lettering. (C. Corn via R. Pickett)

This L-19A (51-7445) is being loaded with a supply bundle on an S-2 bomb rack for a series of operational tests. The outboard bomb rack appears to be natural metal with the release control wiring running under the wing from the rack to the cockpit. (R. Pickett)

Six L-19As perform a formation aerial re-supply drop for ground troops as part of a field exercise during the 1950s. Each L-19 could carry two supply bundles on the underwing shackles. Such methods of resupply could only be carried out if enemy opposition was light. (R. Pickett)

This L-19A (53-8033) is painted in a special high visibility Red and White scheme used when the aircraft was operated over desert or jungle areas. The scheme was used to make the aircraft more easily spotted from the air in case of an emergency landing. (D. Menard)

A Marine OE-1 (136911) of the 2nd Marine Air Wing, Fleet Marine Force shares the ramp at Marine Corps Air Station Cherry Point, NC with F9F Panther and F3D Skynight fighters, The aircraft is painted overall Olive Drab with Black markings. OE-1s were identical to the Army L-19A. (USMC via R. Pickett)

An L-19A (51-12778) of the 3rd Light Aviation Company conducts a tactical photographic reconnaissance mission over Korea. The photographer leaning from the observer's window is using a Fairchild K20 aerial camera. This same type of camera was later pod mounted and carried on the underwing bomb rack. (US Army via C. Yarbrough)

This L-19A (51-12304) was used to test the EDO float installation at Lake Tholocco, Fort Rucker, Alabama. The floats, which were sometimes equipped with wheels for beaching the aircraft, could be installed in place of the conventional landing gear within a few hours. (US Army Aviation Museum)

EDO Twin Float Installation

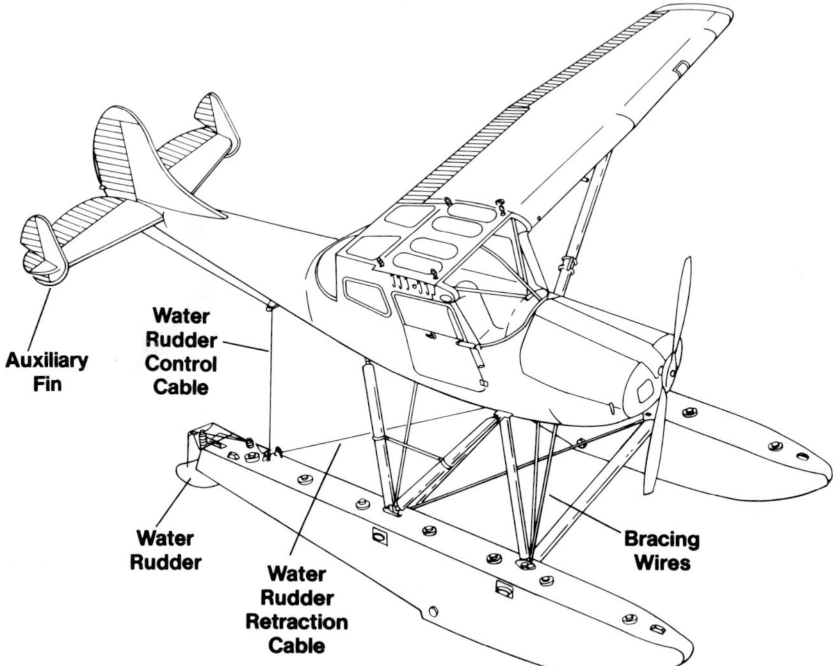

L-19s practice formation flight during training. The L-19 was light on the controls and very responsive, with superb take-off and landing characteristics. Army pilots reported that the L-19 was an excellent aircraft for formation flight and aerobatics. (R. Pickett)

An L-19A (51-7445) flies low to snag the line stretched out by the soldiers on the ground during an aerial message pick-up. After hooking the message, the line would be hauled into the aircraft by the observer. This technique was used in areas where radio silence was being enforced. (R. Pickett)

This L-19A (51-7391) was loaned to Boeing during 1958 to serve as a test bed for the 502-10F turbine engine (later used on the Boeing RP-77 Drone). The insignia on the fuselage side is the Army Aviation Test Board at Cairns Army Airfield, Fort Rucker, Alabama. (US Army Aviation Museum)

Undergoing modification to O-1G configuration, this ex-Texas National Guard O-1A (51-7422) moves along the Cessna assembly line. Modifications included strengthening of the wing hard points, removal of heater assemblies, trainer equipment, and the addition of upgraded radio equipment. (R. Pickett)

(Above) An O-1G (51-12282) of the *Headhunters*, the 219th Light Reconnaissance Airplane Company (RAC) taxies in after a mission at Phu Cat Air Base, South Vietnam on 30 January 1971. The 219th RAC tail emblem on the rudder is Red, White, and Blue. (Norm Taylor)

(Below) *DORIS* an O-1G (51-12785) overflies the Mekong Delta area of South Vietnam during late 1966. Cessna modified ninety-six O-1As drawn from Reserve and Air Guard units to O-1G standards during 1966. The devices under the observer's window are flare canisters. (US Army Aviation Museum)

XL-19B

During 1952 the Boeing Aircraft Company requested the loan of an Army L-19A to test the Boeing 502-8 turbine engine as a possible light aircraft powerplant. The 502-8 turbine engine was a proven engine having been installed in the world's first turbine powered truck and a helicopter. The engine, however, had never been installed in a conventional aircraft.

The Army authorized the test and loaned Boeing one L-19A airframe to serve as an engine test bed. The engine installation was done by Cessna at the company's engineering test facilities in Wichita. The possibility of using the turbine engine on future L-19s led the Army to assign the re-engined L-19A the experimental designation XL-19B.

The Boeing 502-8 turbine engine weighed 250 pounds (some 125 pounds less than the Continental 0-470) and produced 210 horsepower for takeoff and 175 hp for cruise. Being a turbine, the engine was able to operate on any type of fuel from diesel to aviation gasoline. This feature was seen as an advantage for any aircraft that routinely operated with front line troops.

The sole XL-19B (52-1804) takes off on a test flight from the Cessna field at Wichita, Kansas. The XL-19B was the first light aircraft fitted with a turbo-prop engine and set a new altitude record for light aircraft of 37,063 feet during July of 1953 with Cessna test pilot Bill Thompson at the controls. (Boeing)

Inspecting the Boeing 502-8 turbine engine installation of the XL-19B are (Left to Right) Harry McCarter, Cessna Project Engineer; Hank Waring, Cessna Chief Test Pilot; and Sid Shannon, Boeing Turbine Project Engineer. The engine was simple to maintain having no spark plugs or carburetor. (Boeing)

Hank Waring, Cessna's chief test pilot flies the XL-19B over the Kansas country side. Waring reported that flight characteristics of the XL-19B were very similar to the piston engined L-19A, although the engine itself weighed 120 pounds less. (Boeing)

During the flight test phase of the program, it was decided to use the XL-19B in an attempt to set a new world altitude record for light aircraft. The aircraft was specially equipped with an on demand high pressure oxygen system and a barograph to record actual altitudes reached for record purposes. In July of 1953 the test program pilot, Bill Thompson, took off in the XL-19B from the Wichita facility on an altitude record attempt. Gaining altitude steadily, Thompson continued to climb until the XL-19B reached an absolute altitude of 37,063 feet. This new record surpassed the previous record held by a Piper Super Cub of 30,203 feet set over Miami, Florida during 1951 by nearly 7,000 feet.

Although the flight characteristics of the XL-19B were improved over the piston engined L-19A, the Army issued no production contracts for the turbine powered aircraft. Ironically both Italy and France went on to build turbine powered versions of the L-19, while the first turbine aircraft to enter the Army inventory, the Grumman OV-1 Mohawk, was not accepted until 1961. Had the Army opted for the turbine powered XL-19B they could have had a turbine powered aircraft in the inventory some seven years earlier, gaining a tremendous amount of valuable turbine aircraft experience.

The XL-19B test bed was tragically destroyed following engine failure during 1953, ending the short, but illustrious career of the world's first turbine powered light aircraft.

The XL-19B makes a high performance takeoff to demonstrate its takeoff capabilities. Such steep angle takeoffs would be used to clear obstacles, such as trees, normally found around the small fields from which L-19s flew when supporting front line troops. (Boeing)

The XL-19B (left) had a much higher and longer engine cowling with more prominent exhausts than the standard L-19A (right). The turbine engine required installation of enlarged air intakes at the front of the engine cowling. (Boeing)

Engine Cowlings

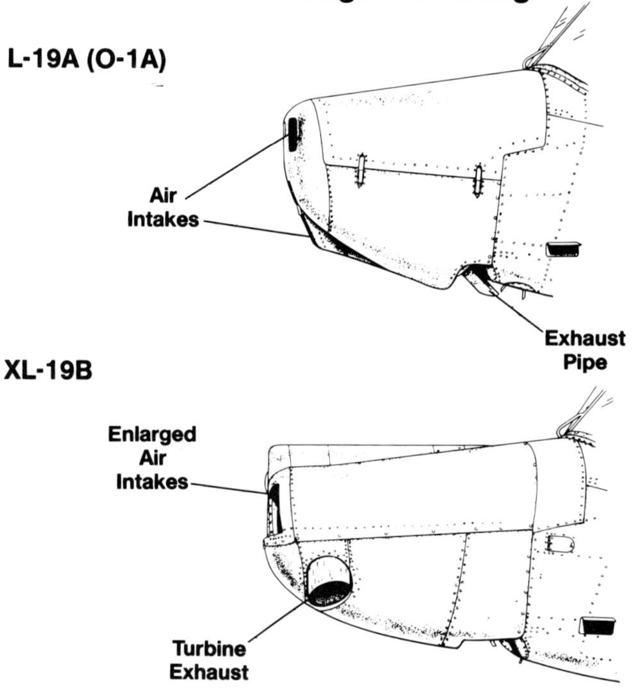

The differences between the Continental O-470 piston engine (left) and the Boeing 502-8 turbine engine (right) are apparent. The turbine engine was a clean installation, without the maze of wiring and conduits required for the piston engine. (Boeing)

Without benefit of ear protection, Cessna ground crewmen prepare to start the turbine engine using an APU. The intakes fitted to the front of the cowling fed air into the engine and were much larger than the cooling intakes on the L-19A cowling. (Boeing)

The XL-19B undergoes ground static tests of the turbine engine. The auxiliary power unit (APU), which plugged into the ground starter plug, was required to turn the turbine up to the proper RPM prior to starting the engine. The piston engine did not require the use of an APU. (Boeing)

The sole XL-19B was destroyed in an accident during 1953. The flaps are fully deployed and it appears that the engine was not running when the aircraft touched down, since only one propeller blade is bent back. Boeing failed to win a production contract for the XL-19B and the project was cancelled. (Boeing via R. Pickett)

XL-19C

The US Army, in an effort to expand their working knowledge of turbine powered aircraft, provided Continental Motors of Toledo, Ohio with two L-19A airframes (52-6311 and 52-6312) to serve as test beds for small turbine engines. Continental, well known for their piston engines, entered the gas turbine engine field during 1952 when they acquired exclusive US rights to produce and develop a line of small turbine engines from the French Turbomeca company.

The turbine chosen for installation in the two test beds, re-designated XL-19Cs, was the XT-51-T-1 (USAF designation XL-51) *Artouste*, a turbine engine not unlike the earlier Boeing engine installed in the XL-19B. The XL-19Cs began a series of test flights during 1954 that revealed that the flight characteristics of the turbine powered aircraft were very similar to the standard L-19A with no major improvement in performance. As a result of these tests, the Army decided against production of the XL-19C.

With the conclusion of the test program, both XL-19C aircraft had the turbine engines removed and replaced by standard piston engines. The aircraft were then returned to the Army inventory as standard L-19As.

The Italians, however, thought the idea of a turbine powered light aircraft was a sound concept and began design work on a turbine powered L-19 in January of 1969. The SIAI-Marchetti Company of Varese, Italy, developed a turbine powered light aircraft based on the L-19, using a 317 hp Allison 250-B15G engine driving a Hartzell three bladed propeller, under the designation SM-1019. The prototype was successfully tested and the Italian Army issued a contract for a total of one hundred aircraft for service in Italian Army Light Aviation units.

One of the two XL-19Cs (52-6311) built prepares for takeoff in a field beside the Cessna plant. The close cowled Continental XT-51-T-1 turbine was similar in performance to the Boeing 502-8 turbine engine used on the XL-19B. (Cessna via R. Pickett)

The engine cowling for the Continental Teledyne *Artouste* turbine engine was about two feet longer than the L-19A cowling to accommodate the longer Teledyne engine. Both XL-19Cs (52-6311 and 52-6312) reverted to L-19A standard after the program ended and were returned to the Army. (Teledyne-CAE via A. Bell)

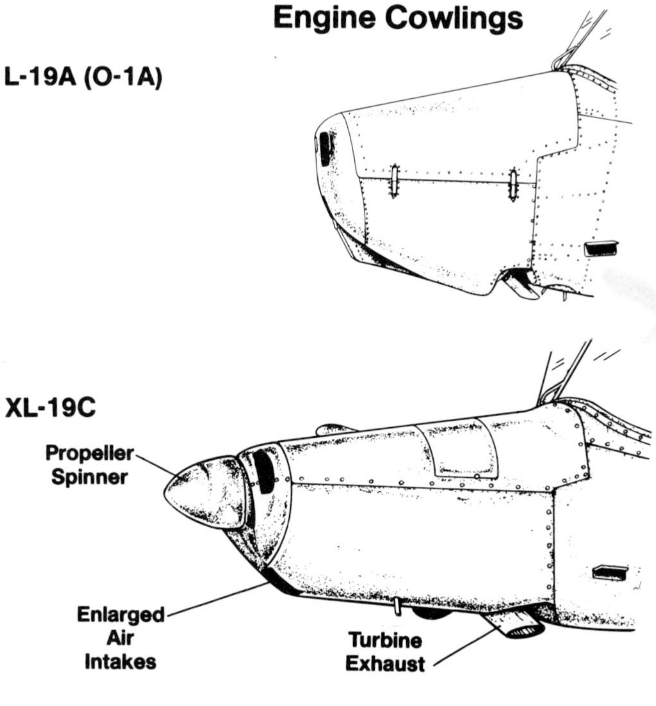

TL-19D
(TO-1D, O-1D, O-1F)

The TL-19D was developed to fill a need within the Army training system for a light aircraft capable of training Army pilots in instrument flight and in the use of a constant speed propeller. The first TL-19D was produced in 1955 with production continuing through 1957. TL-19Ds were assigned to Army units as two seat instrument and proficiency trainers with a full instrument panel being installed in the observer's cockpit. These same aircraft were re-designated during September of 1962, under the Uniform Aircraft Designation System, becoming the TO-1D.

The TL-19D differed from the L-19A primarily in the powerplant and propeller. TL-19Ds were powered by a Continental O-470-15 engine driving a two blade McCauley constant speed propeller in place of the fixed pitch propeller of the L-19A. With the installation of the constant speed propeller gross weight of the TL-19D was increased from 2,200 pounds to 2,400 pounds. Other differences included an adjustable rear seat, a reinforced airframe, deletion of the underwing bomb racks, and relocation of the electrical system circuit breaker panel to the port cabin wall. The relocation of the circuit breaker panel allowed more navigational instruments to be added to the instrument panel for the navigational/instrument training role. Cessna produced a total of 310 TL-19Ds, with all aircraft going to the US Army.

During the mid-1960s the Army training program underwent a series of changes. Fixed wing training declined with helicopter training becoming increasingly more important. During this same time frame, the conflict in Vietnam began to escalate and the need for additional O-1s in Vietnam became apparent. As a result, the majority of TL-19Ds in training units were withdrawn from service and returned to Cessna for modifications to outfit them for combat duty in Vietnam.

Under the modification program the rear instrument panel and cockpit heater were removed and two underwing bomb racks were installed under each wing giving the aircraft the capability of carrying up to eight smoke rockets. The radio equipment was upgraded with the addition of a UHF radio for communications with tactical aircraft. After completion of the modification program the aircraft were re-designated as O-1Ds. To meet an urgent Air Force requirement for FAC aircraft, twenty-two aircraft were loaned to the Air Force under the designation O-1F.

The first aircraft to complete the modification program were assigned to the 73rd Aviation Company, under the command of MAJ John S. Kark. The unit deployed to Vietnam with thirty-three O-1Ds during May of 1963. The aircraft disassembled and shipped by sea and upon arrival in Vietnam were re-assembled by Air Vietnam personnel at Ton Son Nhut Air Base near Saigon. The 73rd Light Aircraft Surveillance Company flew over 30,000 combat hours during their first year in Vietnam, performing duties such as aerial re-supply of forward troops, medical evacuation, flare drops, reconnaissance, forward air control, artillery spotting and convoy escort. During this year the O-1Ds had an average mission ready rate of some 93 percent, a credit not only to the construction of the aircraft but also to the hard work of the Army maintenance crews who kept the Bird Dogs combat ready.

During the late 1960s the US Air Force took over the Forward Air Control (FAC) role in Vietnam and the Army turned over a number of O-1Ds to Air Force control. These aircraft operated alongside the earlier O-1Fs which now became permanently part of the Air Force. Army units continued to operate a small number of O-1Ds until 1972 when the US withdrew from Vietnam and the surviving aircraft were turned over to the Vietnamese Air Force.

This TL-19D (57-2820) of the New York National Guard was later recalled by the Army and modified for duty in Vietnam. The sideward folding rear instrument training panel can be seen through the observer's window. (David Menard)

Front Instrument Panel

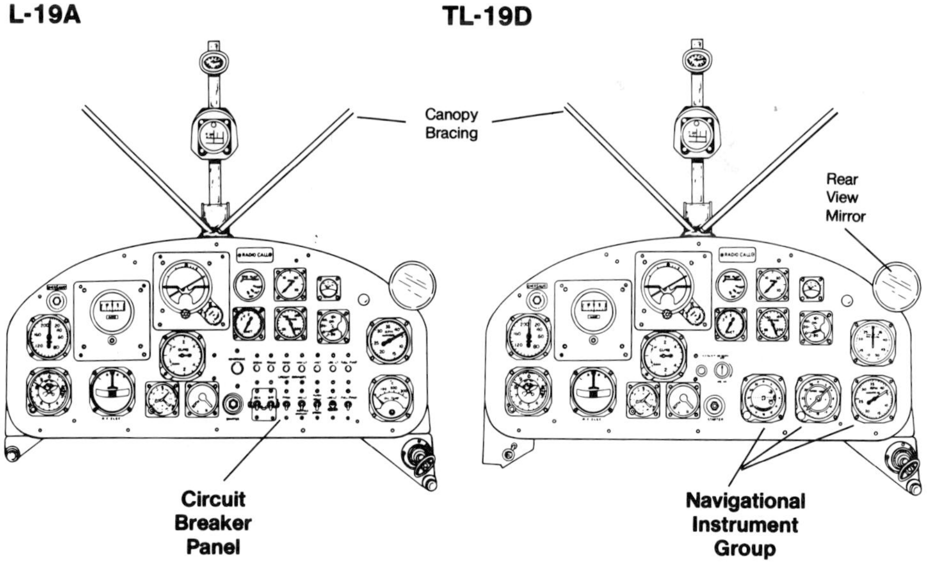

Painted overall International Orange with Black markings, this new production TO-1D (57-2774) was scheduled to serve in the trainer role in the Greenland area, however, it was recalled for modification (like over half of all TL-19Ds produced) and assigned to duty in South Vietnam. (R. Pickett)

Can Tho South Vietnam was home base for this O-1D (57-2902) of the 73rd Aviation Company during 1963. The 73rd AC insignia is carried on the wheel hub and fin tip while Orange identification stripes are carried on the wing uppersurfaces along with White painted flaps. (Charles Blackman)

A newly converted O-1D on an acceptance flight during 1965. Cessna converted a number of TO-1D aircraft to O-1D standards by removing all training equipment and adding four under wing bomb racks, VHF blade antennas, and dual FM radios/antennas. (Cessna)

Rear Folding Instrument Panel

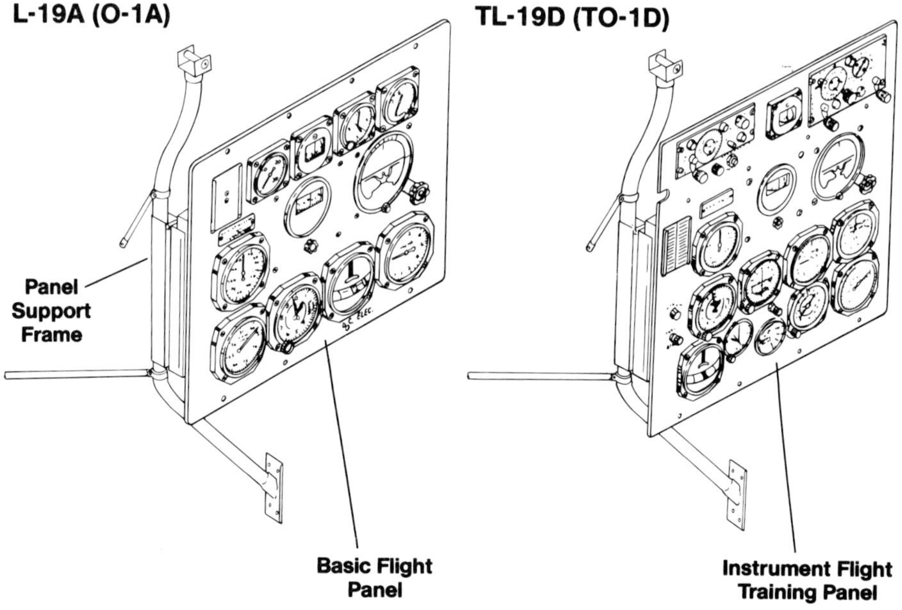

The natural metal Model 305 prototype was pitted against designs from Piper, Taylorcraft and TEMCO for the army light observation aircraft contract in March of 1950.

This overall Olive Drab L-19A of the 8th Army was used to transport President elect Dwight D. Eisenhower on an inspection trip of Korea in December of 1952.

The XL-19B mated the L-19 airframe with a Boeing turboprop engine. The aircraft set a new world's altitude record for light aircraft of 37,063 feet in July of 1953.

This TO-1D carries the high visibility Gloss White and Red scheme used for aircraft that operated over the Arctic or jungle areas.

MAJ Boung used this VNAF O-1A to escape from South Vietnam with his wife and five children. MAJ Boung made the first landing by an O-1 on an aircraft carrier, landing on USS MIDWAY on 30 April 1975.

This TO-1D instrument trainer carries the overall International Orange high visibility scheme used by the Army when operating over the Arctic ice cap in Alaska.

This shark mouthed O-1D of the 19th TASS was one of the first Air Force Bird Dogs to deploy to Vietnam during July of 1963.

The Japanese Land Self Defense Force operates a number of L-19As (O-1As) assembled in Japan by Fuji Heavy Industries during the late 1950s.

A Field Green Drab over White, Marine Corps OE-2 (O-1C) of VMO-6 based at Chu Lai, South Vietnam during the early 1960s.

A number of Air Force O-1s carried three color camouflage schemes that varied from aircraft to aircraft. The Orange identification panels on the wing uppersurface helped attack aircraft spot the O-1.

An O-1D (57-2795) of the 73rd Aviation Company sits in the sun at Ban Me Thuot, South Vietnam during 1963. The 73rd AC served in western II Corps supporting Special Forces camps and the Vietnamese Province Chief. The wing uppersurfaces have an Orange stripe and White flaps to aid identification from above. (Author)

The VHF antenna installation on the fuselage spine was one of the modifications made to TO-1Ds (TL-19D) when they were converted to O-1D standard. These O-1Ds are lined up for a display in honor of GEN Joe Stillwell at Nha Trang, South Vietnam, during March of 1964. Of the 310 TO-1Ds converted, over one third were destroyed. (Author)

Bomb Racks

L-19A (O-1A)

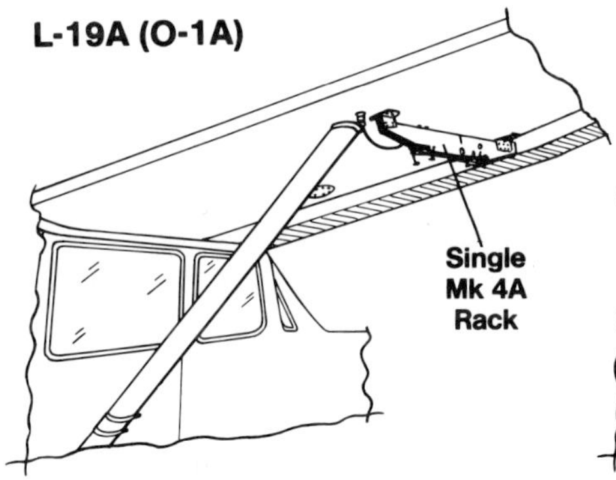

TL-19D (O-1D)

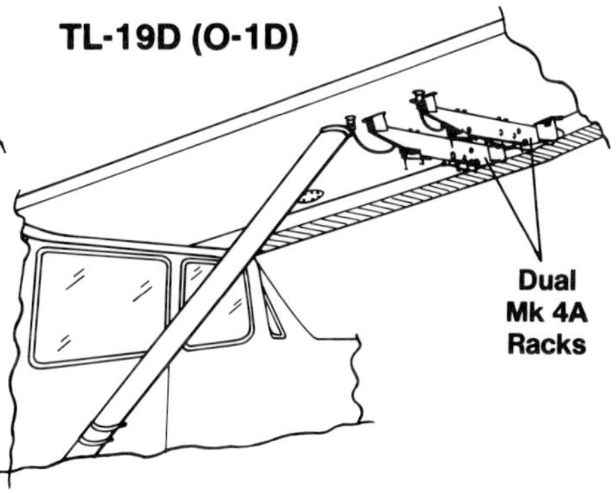

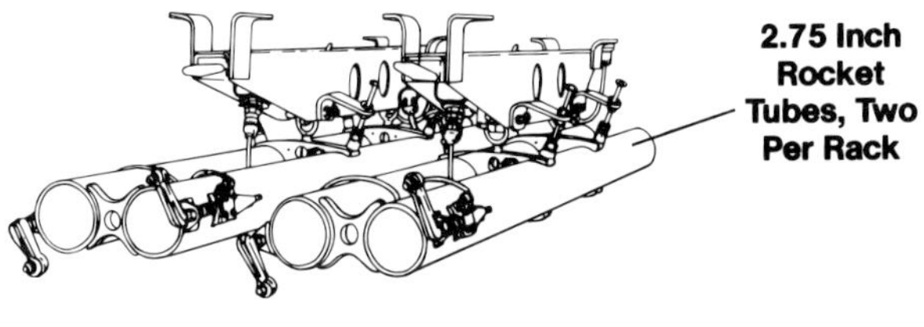

An O-1D (57-2774) overflys the Mekong Delta in IV Corps area while returning from a mission during July of 1963. The national insignia on the fuselage side presented the VC with a convenient aiming point and was usually painted out. (Author)

The primary photo reconnaissance aircraft employed by the 73rd Aviation Company was this O-1D (57-2901) damaged by a ground loop at Nha Trang, South Vietnam during 1963. This aircraft still carries the national insignia on the fuselage side and has not had the White and Orange upper wing identification panels painted on. (Author)

A maintenance crew of the 73rd Aviation Company paint Orange panels on the wing uppersurface of an O-1D at Nha Trang during 1963. O-1Ds normally operated at low levels and in order to be seen by attack aircraft, Orange and White identification panels were painted on the wing uppersurfaces. (Author)

Repaired and once again flying photo reconnaissance missions over South Vietnam, O-1D 57-2901 was later destroyed while on a combat mission. The aircraft carries two KA-39 camera pods on the underwing pylons. On night missions flares would be carried in place of one of the camera pods. (Author)

Antenna Configuration

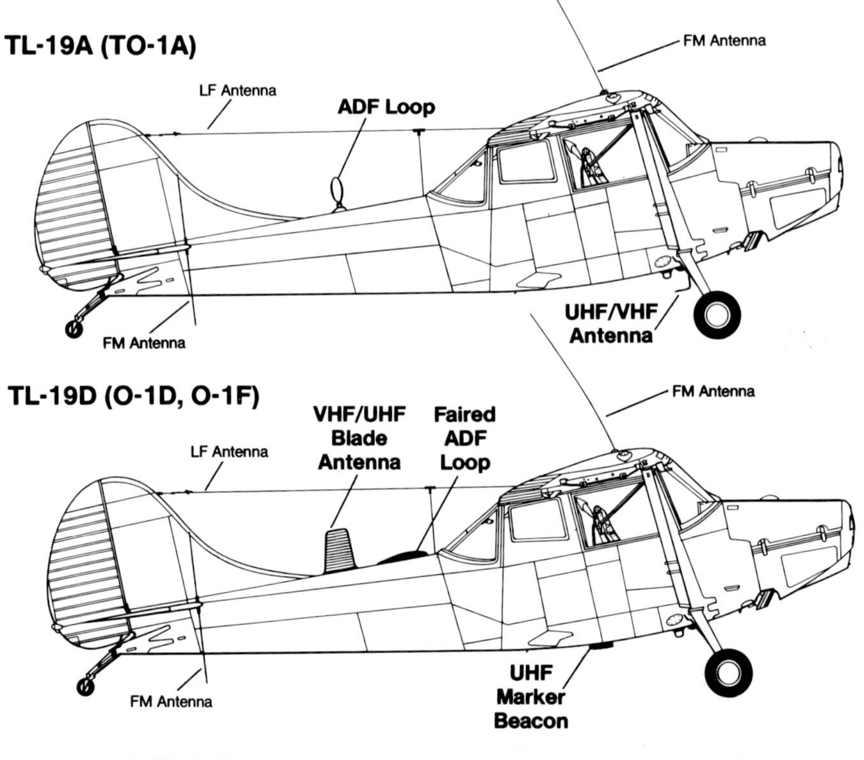

This O-1D (57-2880) undergoes a 100 hour maintenance inspection at Nha Trang during 1964. The inspection usually required two days to complete and consisted of inspection, repair, cleaning, and replacement of damaged components. (Author)

Propeller

L-19A (O-1A)

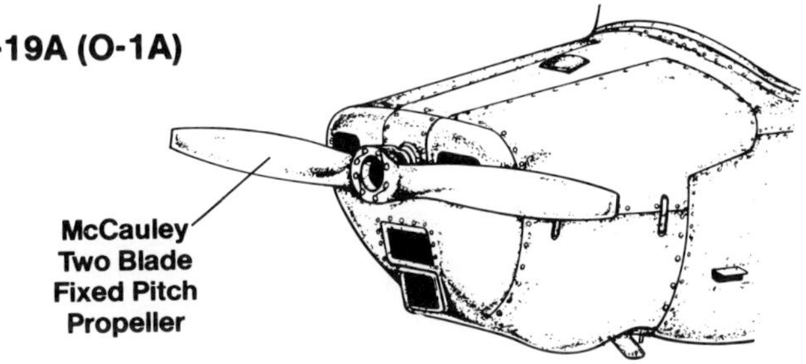

TL-19D (O-1D)

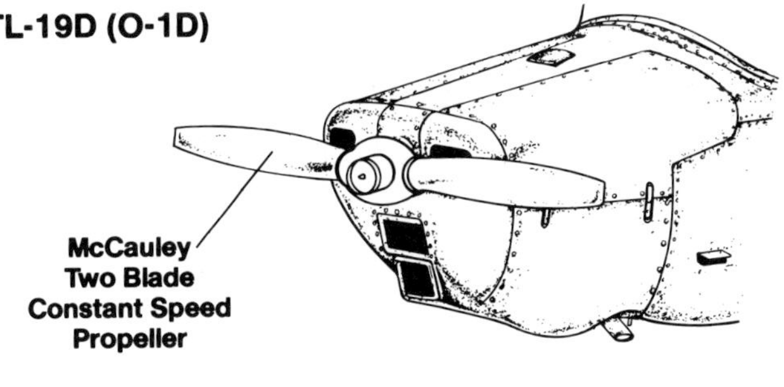

After every twenty-five hours of flight a mini-inspection, oil, and filter change was performed. This O-1D (57-2880) of the 73rd Aviation Company is undergoing a twenty-five hour check at Ban Me Thuot during 1964 where conditions were even more austere than at the 73rd's home base of Nha Trang. (Author)

Aircraft serving with the 73rd Aviation Company in South Vietnam had the company insignia painted on the vertical stabilizer. This Vietnamese artist painted the company insignia on thirty-three aircraft during the unit's stay at Nha Trang Airfield. (Author)

An O-1D (57-2880) takes off from Ban Me Thuot during 1964 with a load of cargo for a Special Forces camp on the underwing bomb racks. The O-1D performed a variety of tasks over Vietnam from aerial re-supply to target marking. (Author)

An O-1D drops a pair of cargo bundles over the drop zone at a Special Forces Camp in South Vietnam. Each wing pylon could carry up to 125 pounds of cargo for a combined weight of 500 pounds. O-1s seldom carried this maximum load since the weight and drag severely reduced performance. (Author's collection)

Parked in its revetment at Can Tho Vietnam this pair of Army O-1Ds await their next mission during 1968. Army O-1Ds were used to mark targets for VNAF fighter-bombers and usually carried a Vietnamese observer. The aircraft are painted in a subdued scheme of Olive Drab with Black lettering. (C. Blackman)

When the Army turned over a number of O-1Ds to the Air Force, they were repainted in USAF colors. This O-1D (57-2897) of the 19 TASS at Bien Hoa, flown on missions in January of 1966, was partially repainted with the front half in USAF colors and the rear half in the Army colors. (D. Menard)

This O-1F (57-2926) of the 117th Aviation Company was painted in an experimental camouflage scheme of Olive Drab and Tan over Light Gray during 1964 and was one of fifteen TL-19D aircraft modified to O-1F standards by Cessna during 1962. (Author)

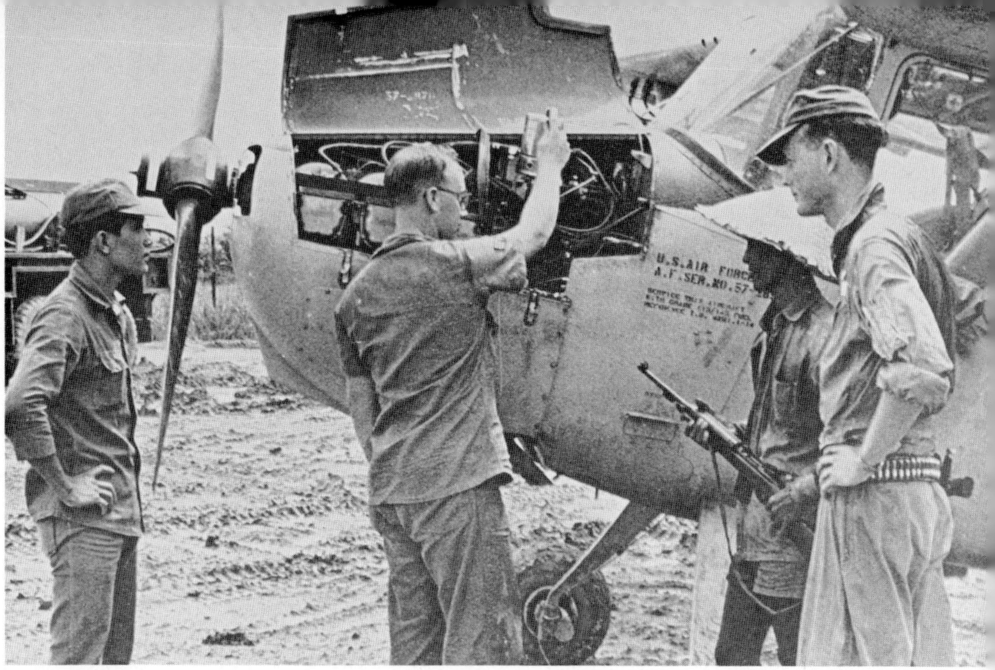

Under the watchful eye of a local Vietnamese guard, an Air Force ground crewman adds oil to the engine of an O-1D (57-2870) being prepared for a mission over South Vietnam. Despite these precautions the aircraft was later destroyed. (USAF via H. Morgan)

Ronnie's Racer **an overall Light Gray O-1F of the 21st TAAS is parked and chocked in a protective revetment at Nha Trang during December of 1970. These revetments were designed to protect the aircraft from fragments during VC rocket attacks. (N.E. Taylor via Menard)**

L-19E (O-1E)

The L-19E was basically an improved L-19A, incorporating the reinforced airframe, wing, and four underwing bomb racks of the TL-19D. The E model was powered by the 213 hp Continental 0-470-11 engine with a two blade fixed pitch McCauley propeller in place of the constant speed propeller used on the TL-19D. The L-19E was the second most widely produced Bird Dog variant with production beginning during 1956 and continuing through 1957 for a total of 469 aircraft.

The build up in Vietnam and the subsequent strengthening of the Army Aviation Program prompted the Army to issue additional production contracts for Bird Dogs. To meet these contracts the Cessna production line was re-opened during 1961 with production of Bird Dogs continuing through 1963. During this period, Cessna's Military Division built an additional 97 aircraft bringing the total L-19E production to 503 aircraft.

In September of 1962 the L-19E designation was changed to O-1E, conforming with Department of Defense directives standardizing aircraft designations. The O-1E was widely used, serving in a number of roles including both combat operations and the basic and instrument trainer role when fitted with a complete navigational instrument panel in the observer's cockpit.

The O-1E like the O-1A, D and F models served in Vietnam with both the Army and Air Force. Before deployment to Vietnam O-1Es were brought up to current combat standards by removing all trainer features and installing additional radio equipment, self-sealing fuel tanks, and an armored pilot's seat (which was also fitted to earlier O-1 models in the field). These modifications brought the Bird Dogs weight up to 2,400 pounds but were necessary to handle the increasingly demanding combat environment in South East Asia. The O-1E was the mainstay of Air Force Forward Air Control units until the appearance of the Cessna O-2 Super Skymaster and Rockwell OV-10 Bronco. Both of these aircraft were larger, twin engined aircraft capable of carrying weapons for self-defense and light strike.

Vietnam was the last action for the O-1 in US Army service. Helicopters were rapidly replacing light aircraft in Army units and by the mid-1970s the majority of O-1s left in the Army inventory were turned over to either the Civil Air Patrol or Army National Guard Units.

The crew of this Army Reserve O-1E (56-2613) conducts preflight checks before another pilot training mission. As the war in Vietnam progressed most O-1s in service with the Guard and Reserve were recalled for combat duty with the Air Force. (D. Menard)

Front Instrument Panel

TL-19D (O-1D) **L-19E (O-1E)**

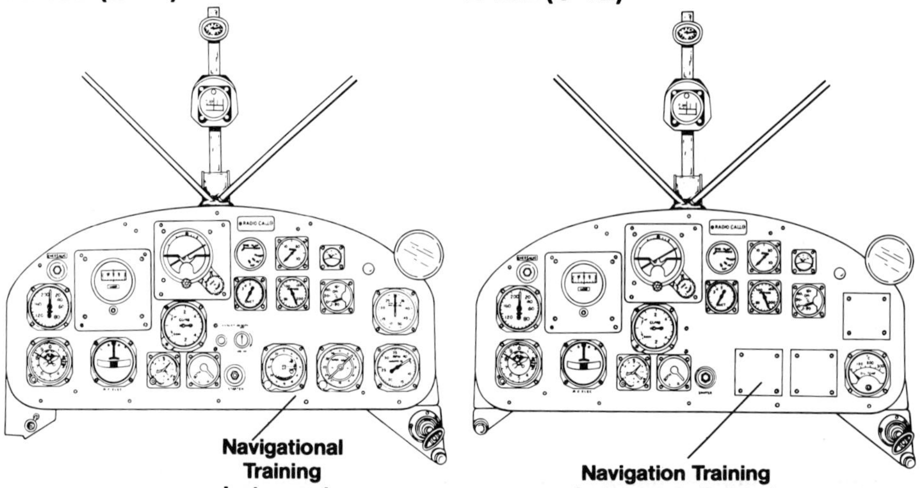

A new production L-19E (O-1E) on a acceptance flight over the Kansas country side during 1956. Used in the trainer role until 1967, this O-1E (56-2647) was modified by Cessna under an Air Force contract that called for removal of all training equipment and upgrading the radios to current combat standards. (Cessna via R. Pickett)

Prior to its modification during 1966, this Army O-1E (56-2691) was utilized in the trainer role with Orange panels on the nose, rear fuselage, fin, and wing to warn other aircraft that a student was on board. (US Army Military History Institute)

The instrument panel of the O-1E was similar to the TO-1D with the exception of the removal of the manifold pressure gauge, Azimuth indicator, and course indicator selector gauge. The panel was supported by rubber bushings to dampen engine vibration. The box between the rudder pedals contains the 24 volt battery. (R. Pickett)

An Army O-1E (56-2650) takes off from an unimproved field during 1965. The O-1E was an improved O-1A having the reinforced airframe, strengthened wing, and provisions for four underwing bomb racks of the O-1D. The O-1E was the second most widely produced variant of the Bird Dog. (R Pickett)

Painted in overall trainer Orange with Black anti-glare panel this O-1E (56-2661) takes off from an Army airfield. This aircraft, like ninety-five other Army O-1Es, was passed to the Air Force during 1966. (US Army Military History Institute)

This shark mouthed O-1E (56-2688) of the 183rd Aviation Company undergoes preflight before a mission over the II Corps tactical zone in December of 1966. The 183rd operated in the light surveillance role over South Vietnam from 1966 to 1971. (Terry Love via R. Pickett)

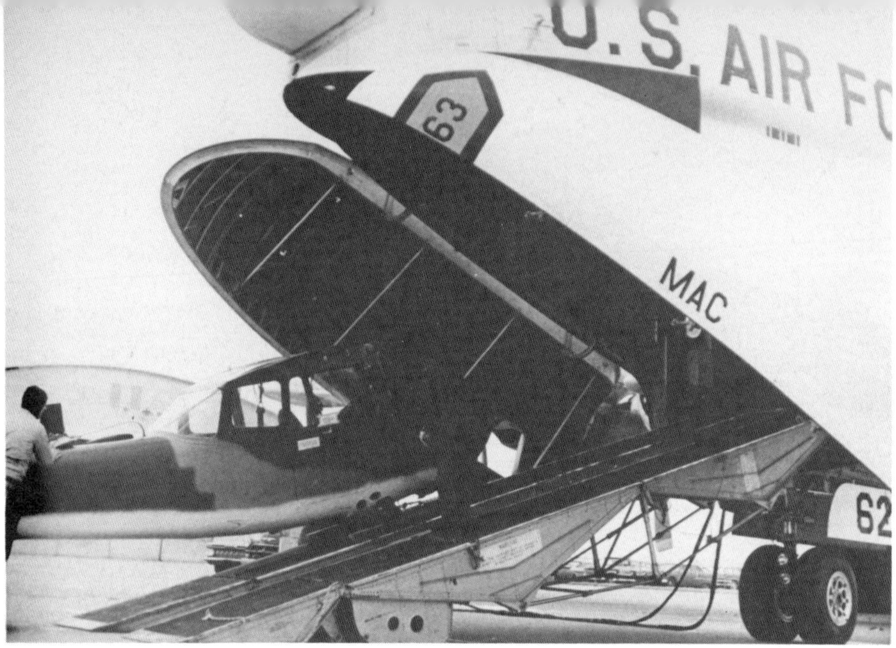

A camouflaged O-1E is loaded on board a Douglas C-124 Globemaster for shipment to Vietnam. During 1966 Cessna modified a total of eighty-eight early model Bird Dogs to meet current combat standards under a $687,000 USAF contract which included electronic and airframe upgrades. (R. Pickett)

Before pilots deployed to Vietnam they went through a training course in Forward Air Control (FAC) techniques under the 4410th Combat Crew Training Squadron. This O-1E (56-4232) of the 4410 CCTS tied down on the ramp at Holly Field, Florida on 2 June 1969 carries Red and White trainer bands on the fuselage side. (Norm Taylor)

Antenna Configurations

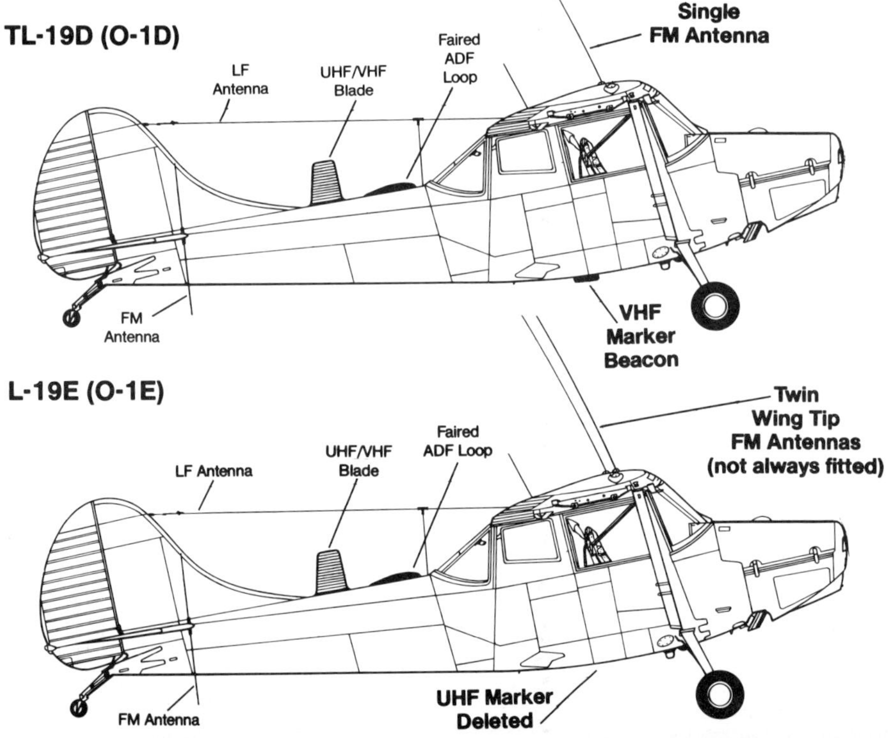

This FAC O-1E (56-2602) is flying in his assigned area searching for any change in the land that might indicate the presence of the Viet Cong. If a target is found the FAC will mark it with WP rockets so fighters can spot the enemy positions. (USAF via H. Morgan)

Pilot's Seat

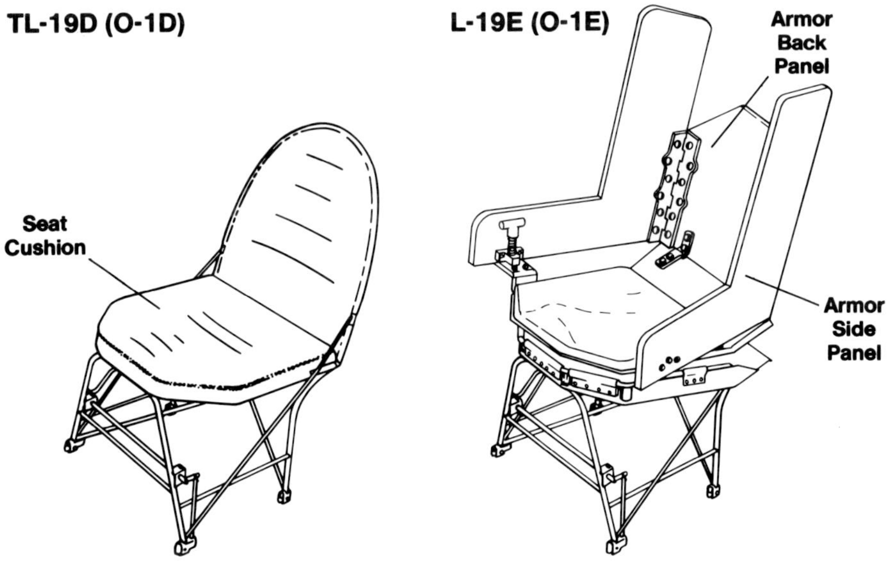

Operating with the 21st TASS out of Nha Trang this O-1E (57-6026) flies over the central highlands of South Vietnam during March of 1966. The White panel under the 'Air' in the fuselage marking is the armament panel which was marked with the type of armament carried on the aircraft. (USAF via C. Yarbrough)

OE-2 (O-1C) BIRD DOG II

The Marine Corps first became acquainted with the Bird Dog during 1951 when a number of L-19A were acquired for use by Marine Observation Squadrons One and Six (VMO-1 and 6) in Korea under the designation OE-1 (later re-designated O-1Bs.) A total of 101 OE-1s would be acquired by the Marines between 1951 and 1967, some being ex-Army or Air Force aircraft. During the mid-1950s Marine planners issued a requirement for an observation aircraft with more power and greater range. Cessna responded to the requirement with the Model 321 a much improved L-19.

The Model 321, although based on experience gained with the L-19/O-1 series, was a completely new design and would be the first liaison aircraft specifically designed from the outset with provision for target marking rockets built in. The Model 321 used the tail assembly and thirty-six foot wings of the Cessna Model 180 coupled with a new strengthened fuselage. The prototype was built to handle optional powerplants and could be powered by one of two different engines, the 260 hp Continental 0-470-2 or a 260 hp supercharged Continental 0-470-2 engine.

To meet Marine range requirements the fuel capacity was increased to fifty-two gallons, with the fuel tanks being the self-sealing type for protection against ground fire. The additional fuel gave the prototype a range of 708 miles. Performance of the Model 321 was impressive with a top speed of 185 mph, a cruising speed of 135 mph, and a maximum dive speed of 220 mph. Service ceiling was over 26,000 feet when fitted with the supercharged Continental engine. The empty weight of the prototype was 1,830 pounds, with a gross weight of 2,650 pounds.

Additionally, the prototype featured armor protection for the crew in the form of a nylon weave flak curtain that extended from the window sill to the cockpit floor. Removable three-quarter inch armor plate was also fitted under the front and rear seat cushions for protection from small arms fire from below. Offensive armament consisted of four underwing pylons for rockets, gun pods, or bombs up to 250 pounds. After successfully completing acceptance tests, the prototype was ordered into production by the Marine Corps under the designation OE-2. Marine procurement totaled twenty-seven aircraft, twenty-five built during 1955 (140078-140102), and two built during 1956 (148250-148251). Additional procurement was curtailed because of the higher costs of the OE-2 which prevented its use on a wider scale.

The Marines used the OE-2 (re-designated the O-1C during 1962) in South Vietnam mainly in the Da Nang area during the 1960s as a Forward Air Control (FAC) aircraft directing close air support aircraft against enemy units in direct contact with Marine ground forces. In addition OE-2s served in other roles such as liaison, artillery target marking, reconnaissance, and aerial re-supply drops. Marine Observation Squadrons Two and Six (VMO-2 and VMO-6) operated the OE-2/O-1C, O-1B and O-1E alongside UH-1 Huey helicopters until the Fall of 1969 when the Bird Dogs were replaced by Rockwell OV-10 Broncos.

OE-2/O-1Cs carried two different color schemes during their service with the Marine Corps. Originally the aircraft were delivered in Dark Olive Green uppersurfaces with White undersurface. In June of 1961 the US Navy issued a directive (MIL-C-18263 FAS) calling for a change in the camouflage to an overall Nonspecular Field Green scheme. OE-2/O-1Cs continued to serve with the Marine Corps until March of 1970 when the last Bird Dogs were retired from Marine units.

The fourth production OE-2 (140081) awaits delivery to a Marine unit outside the Cessna plant during 1955. This aircraft later saw service with the Marines in Vietnam, being based out of Chu Lai during the 1960s. This OE-2 was finished in the standard Marine scheme of Olive Drab uppersurfaces with Gloss White undersurfaces. (Cessna via R. Pickett)

Engine Cowling

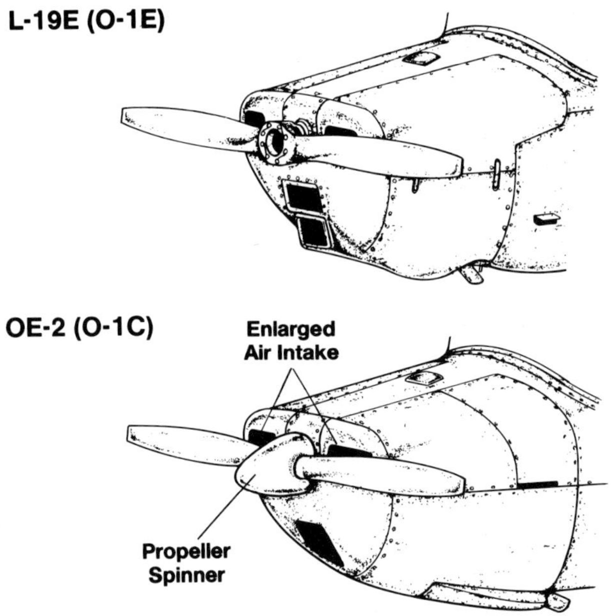

An early production Marine OE-2 prepares for takeoff from a grass field. Marine OE-2s operated alongside earlier OE-1s in Marine Observation Squadrons Two and Six (VMO-2 and VMO-6).

This OE-2 (140078) was accepted by the US Marine Corps during 1955. The OE-2 was the first liaison aircraft to be designed and produced with built-in fittings for underwing pylons. Powered by a Continental O-470-2 supercharged engine the OE-2 had a top speed of 185 mph. (R. Pickett)

Fin and Rudder

L-19E (O-1E)

OE-2 (O-1C)

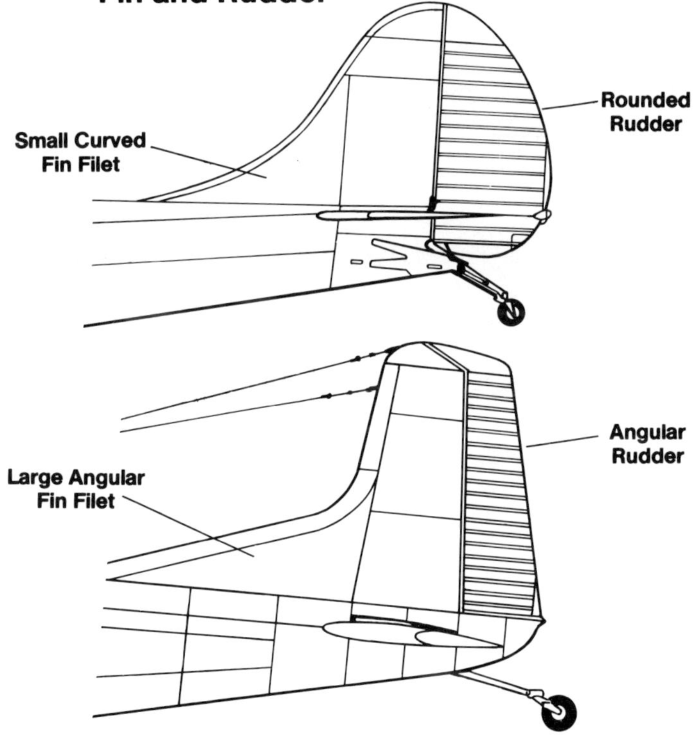

The OE-2 was built exclusively for the US Marines and was delivered from the factory in Flat Olive Drab uppersurfaces over White undersurfaces. The OE-2 was capable of carrying heavier underwing loads (up to 250 pounds) than the earlier OE-1. (R Pickett)

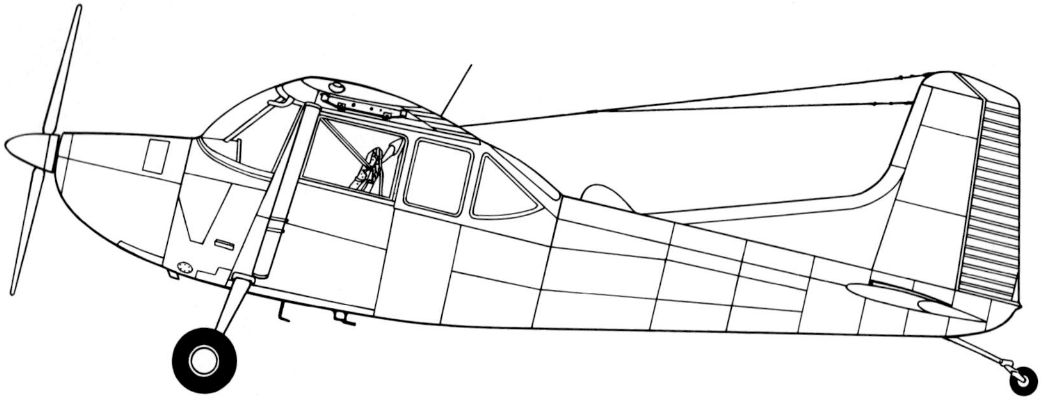

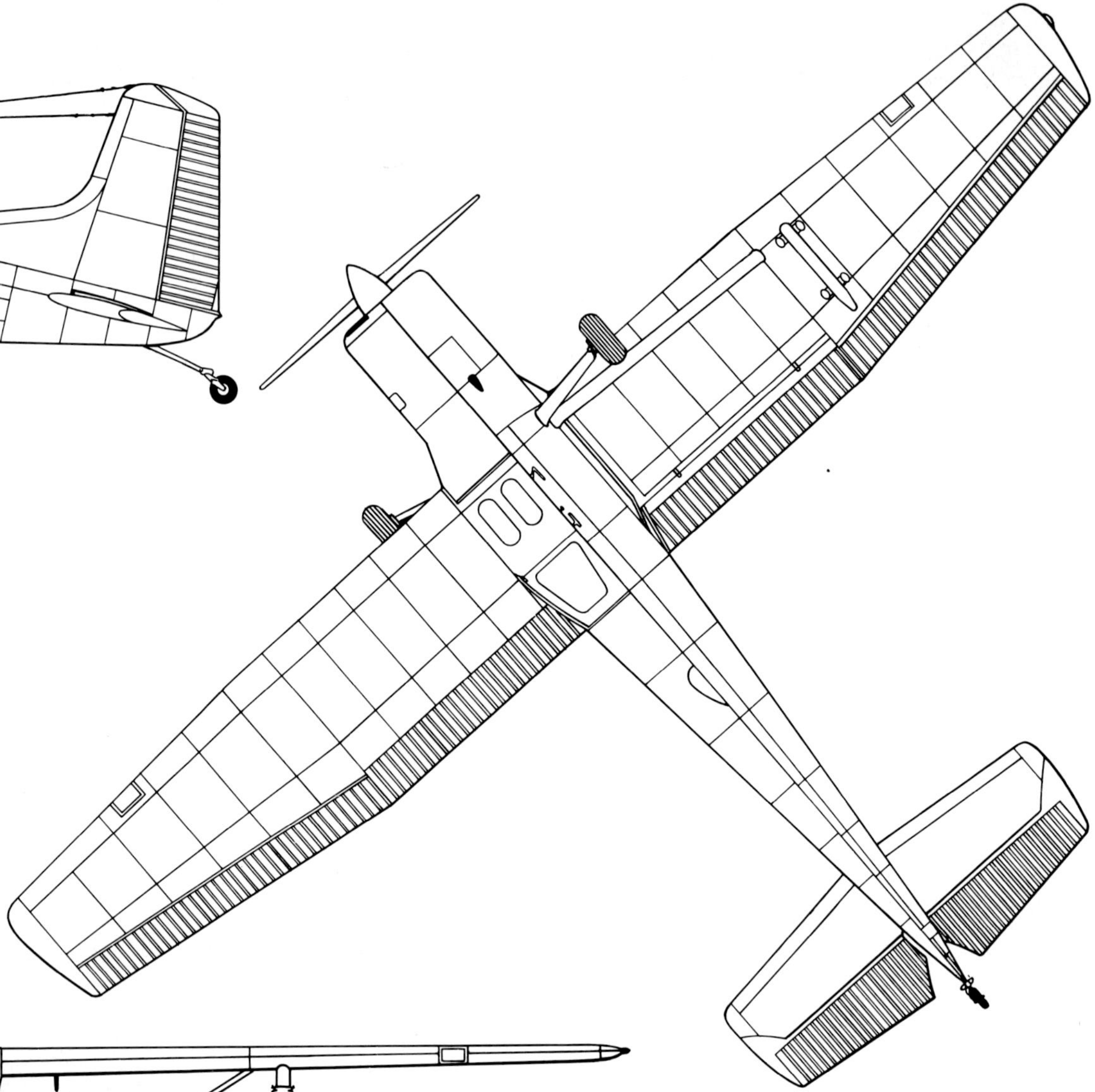

Specifications

Cessna OE-2 (O-1C) Bird Dog II

Wingspan	36 feet
Length	25 feet 9 inches
Height	9 feet 1 inch
Empty Weight	1,830 pounds
Maximum Weight	2,650 pounds
Powerplant	One 265 hp Continental O-470-2 six cylinder supercharged engine
Armament	Four 2.75 inch smoke rockets
Performance	
Maximum Speed	185 mph
Service ceiling	26,200 feet
Range	708 miles
Crew	Two

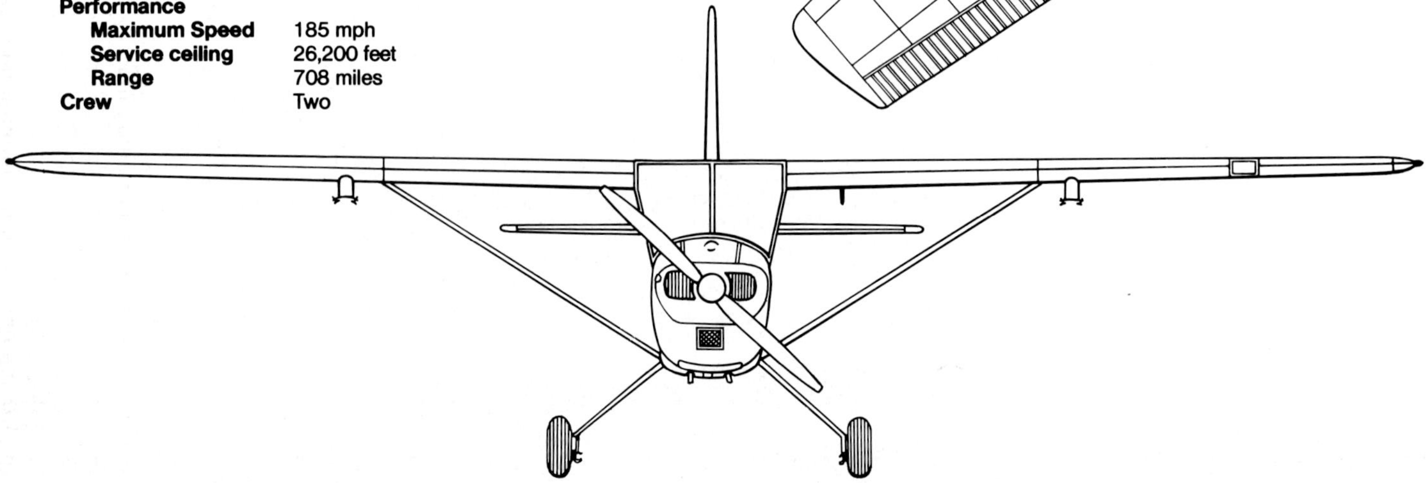

Whiskey Bravo 1, an OE-2 (140082) of Marine Observation Squadron Six (VMO-6), was stationed at Chu Lai along with a number of OE-1s and UH-1E Huey helicopters. The OE-2 was produced as a replacement for the OE-1, however, its increased cost prevented wider use. (Cessna via C. Yarbrough)

VMO-2 and VMO-6 were mixed squadrons operating both OE-2s and the earlier OE-1(O-1B) Bird Dog. This OE-1 carries the overall Flat Field Green Drab camouflage adopted for Marine observation aircraft in June of 1961.

In 1962 all military aircraft designations were changed and the OE-2 became the O-1C. This O-1C (140094), parked beside a USAF O-1E (56-4186) during the late 1960s, has been repainted in overall Flat Field Green Drab with White markings. (R. Pickett)

The O-1C was designed from the onset with flak curtains and armored seats for crew protection. This O-1C (140100) finished in overall Flat Field Green Drab with the wing tips, horizontal stabilizer and fin tip in Orange. All lettering is in White. (US Navy Aviation Museum via W. Johnson)

Forward Air Control (FAC) in Vietnam

The concept of an airborne controller capable of directing both artillery and air support/strike aircraft over the battlefield came into wide spread use during the Second World War. Using different radios, the observer could talk both to ground units and airborne fighter-bombers, dramatically increasing the responsiveness of air support to the ground troops. During the Second World War observation pilots primarily flew the Piper L-4 Grasshopper, which was the mainstay of the US Army Air Corps liaison and observation squadrons.

By the beginning of the Korean War, the mission of the observation pilot had been refined with these pilots now becoming known under the term, Forward Air Controllers or FACs. At the start of the Korean War, the Army was flying Stinson L-5s in the Artillery Observation (AO) and FAC role while the Air Force flew the T-6 Texan. The USAF FAC mission in Korea was carried out by the 6147th Squadron, known as the 'Mosquitos' (named for an early radio call sign.) The squadron was staffed by fighter pilots drawn from operational fighter-bomber units, because there had been no time to specially train pilots for the unique FAC role. While the 'Mosquitos' usually controlled Air Force tactical fighters, they were often called upon to direct Army artillery and Navy/Marine Corps fighter-bombers as well. Coordination between the services, however, was hampered by the radio fit on the T-6s. When the Army and Air Force began to receive L-19A Bird Dogs, this coordination was dramatically improved. The L-19A carried the necessary radio equipment to coordinate between air and ground units, flew low and slow to detect targets, and carried target marking smoke rockets to identify the target for the fighters.

Riding shotgun for a truck convoy this FAC TO-1D (57-2814) searches for possible VC ambush sites along the road. Regularly flying over the same areas ensured that the pilots and observers knew the territory and could detect any change in the normal routine. (USAF via R. Pickett)

FAC pilots sometimes tired of being shot at without being able to shoot back. This Army FAC pilot has hung a pair of seven shot rocket pods loaded with high explosive rockets on the underwing bomb racks of his O-1D to even the odds. (Schwinghanner via Mutza)

The unconventional war in Vietnam quickly showed that close coordination between air and ground units was an absolute necessity. When troops traveled along Vietnamese roads it was vital to have a FAC aircraft in the air to avoid being ambushed. The presence of an airborne FAC would often deter the elusive Viet Cong from setting up an ambush and if one was detected, the FAC could warn the convoy and direct fighter-bombers to the target so that the convoy could safely move on.

Initially the Forward Air Controller mission in Vietnam was undertaken by Army and Vietnamese Air Force (VNAF) L-19/O-1 units. By the mid-1960s, however, the Air Force had taken over the FAC role and continued to run the majority of FAC missions for the duration of the US involvement in Vietnam. The Army retained a number of O-1s primarily for the Artillery Observation (AO) role directing artillery fire from fire support bases and to provide liaison services for the scattered Special Forces camps. Marine Corps O-1 units did provide FAC services for Marine Air Wing strike aircraft operating out of Da Nang in direct support of Marine ground units until replaced by newer aircraft during the late 1960s.

In July of 1963, the Air Force deployed its first FAC squadron, the 19th Tactical Air Support Squadron (TASS) to Bien Hoa Air Base, South Vietnam. This initial deployment was later increased by three additional squadrons, the 20th TASS stationed at Da Nang, the 21st TASS based at Nha Trang, and the 22nd TASS stationed at Bien Thuy. The USAF FAC units in Vietnam, like their counterparts in the Army, amassed an amazing number of flying hours. One unit, the 21st TASS operating out of Nha Trang, flew over 39,000 sorties during 1967 while logging over 50,000 combat flight hours.

A number of O-1s were also involved in clandestine FAC missions along the Ho Chi Minh Trail. Painted Flat Black these O-1s, operating out of Thailand, flew over the 'Trail' at night searching for truck convoys and other signs of North Vietnamese activity. Additionally, these Black painted O-1s were active over Laos and Cambodia supporting CIA and Special Forces units operating in these countries. O-1s used for these clandestine missions were usually Air Force O-1Fs modified with a shielded roof mounted red rotating beacon light that could only be seen from above and a frame holder on the fuselage side into which could be inserted placards with whatever national insignia might fit the mission.

An American FAC pilot and his Vietnamese observer discuss their next mission in front of an overall Light Gray O-1E (56-2610) of the 19 TASS at Bien Hoa during March of 1968. The O-1 carried a striking Red, White and Black 'dog-mouth' painted on the cowling and fuselage sides. (USAF via D. Menard)

The FAC job was dangerous and demanding on both the pilots and their Bird Dogs. The Viet Cong would use every weapon they had to bring down a FAC once they had been spotted, since they realized that the presence of the FAC would bring strike aircraft down on them. Pilots often flew low enough to mark targets with smoke or White phosphorous grenades thrown from the side windows. A number of incidents involving O-1s were unique. The sole aerial victory scored by the Khmer Air Force of Cambodia came at the expense of a USAF FAC O-1 during 1964. The O-1 strayed across the Cambodian border and was shot down by a KAF AT-28A Trojan fighter/trainer, killing the USAF pilot and his Vietnamese observer. One Air Force FAC pilot won the Congressional Medal of Honor for his actions while on a mission over the Central Highlands on 24 February 1967. CAPT Hillard A Wilbanks, flying an O-1G (51-5078) of the 21st TASS, made repeated attacks against a large North Vietnamese Army force firing his M-16 rifle from the side window of the O-1 to disrupt the NVA's planned ambush of a South Vietnamese Ranger unit. He was unfortunately killed on this mission when the VC shot down his aircraft. The USAF lost a total of 175 O-1s to all causes during the Vietnam War, while the Army lost 391, and the Marine Corps lost nine for a total of 575 aircraft.

As time went on the FAC mission became more demanding, bringing additional requirements that were beyond the capabilities of the O-1. The Air Force wanted more range and fire power, while the Marines want higher speed, more firepower, and greater survivability. As a result, a number of newer aircraft began to slowly replace the Bird Dog in the FAC role. Air Force units began to re-equip with the Cessna 0-2A Super Skymaster and Rockwell OV-10A Bronco during 1967, while the Marines transitioned to the OV-10A and also employed TF-9J Cougar and TA-4J Skyhawk as 'Fast FACs' in high risk areas.

The FAC mission was performed by a special breed of dedicated pilots and observers who flew in unarmed or lightly armed aircraft over heavily armed enemy territory. They routinely risked their lives so that all of the available firepower could be brought to bear on the enemy, saving American and South Vietnamese lives on the ground.

This USAF overall light Gray O-1G (51-12824), armed with four White phosphorus (Willie Peter) rockets on the inboard bomb racks, searches for enemy activity over the Central Highlands of South Vietnam. The USAF flew the O-1G, O-1D, and O-1E in Vietnam. (USAF via C. Yarbrough)

An O-1E (56-4200) of the 20th TASS out of Da Nang flies a bomb damage assessment mission over a target area previously hit by fighters. FACs performed both pre-strike and post-strike target assessment missions over Vietnam. (US Army Military History Institute)

A USAF FAC pilot inspects the White phosphorous (WP) rockets on his O-1 during preflight inspection. These rockets were the primary 'armament' of O-1s used in the FAC role. The WP rocket made a highly visible white burst when it hit the target. (USAF)

A pair of O-1Es of the 20th TAASS tied down on the pierced steel planking (PSP) ramp at Da Nang during 1967. Even when secured for routine servicing, the O-1s were normally kept armed and ready for quick reaction missions. (Tom Hanson via Wayne Mutza)

FAC Weapons

2.75 Inch Folding Fin Aerial Rocket

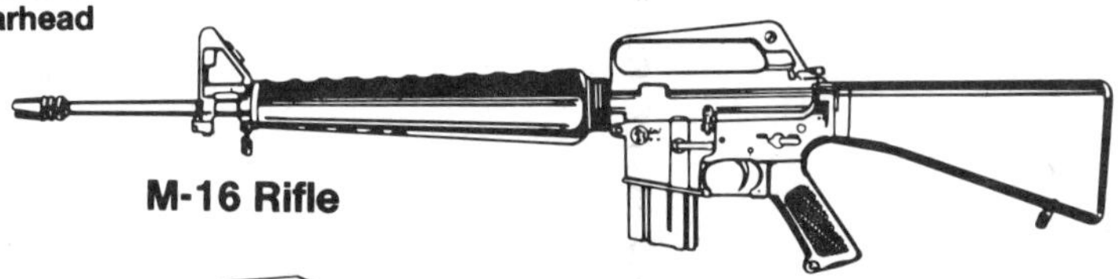

M-16 Rifle

M-185 Smoke Grenade

M-34 White Phosphorous Grenade

Making final adjustments to an M-60 machine gun mounted in the rear seat of his O-1, CAPT Roger Krell prepares to do battle with the VC anti-aircraft gunners during May of 1966. The O-1G (51-11976) was dubbed *Little Puff* and was one of two O-1s modified to test the effectiveness of an armed O-1. (USAF via H Morgan)

An Air Force O-1E enroute to the target area in the Central Highlands flies above the range of VC small arms fire. Once in the target area, the FAC will descend to just above the trees to search for VC activity. (USAF)

This Vietnamese Air Force O-1G (51-12272) based at Phu Cat during 1971 operated in the FAC role alongside USAF units. The Bird Dog is overall Light Gray with anti-glare panel and wing stripes in Black. The 'O' preceding the serial number indicated that the O-1 was over ten years old. (USAF via H. Morgan)

This O-1E (56-2504) is camouflaged in the Tactical Air Command scheme of Tan and Brown uppersurfaces over White undersurfaces. There was no standardized pattern for this camouflage and its application varied from aircraft to aircraft. (Cessna via R. Pickett)

Escape From South Vietnam

On the island of Con Son, off the coast of South Vietnam, Vietnamese Air Force Major Boung, his wife, and five children waited for dawn when they would board the tiny O-1A that would take them to freedom. The fall of Saigon was eminent and MAJ Boung knew that he and his family could not live under a Communist regime. To escape the North Vietnamese he had formulated a desperate plan; to put as much fuel into the tanks of his O-1A as he could and fly from Con Son Island out over the South China Sea in search of an American aircraft carrier.

On 30 April 1975 he put his plan into action. As the situation in Vietnam had become more desperate the US Navy had stationed three aircraft carriers, USS MIDWAY, USS CORAL SEA and USS ENTERPISE, off the South Vietnamese coast. These carriers were part of OPERATION FREQUENT WIND, the aerial evacuation of all US personnel from Vietnam.

Flying out over the South China Sea, MAJ Boung knew that he had only enough fuel for a one hour search. If he did not find a carrier before his fuel was exhausted, he would have to ditch in the sea. Good fortune, however, was riding with them and he soon spotted the carrier MIDWAY. Circling the ship, he wrote a note on a navigational chart and threw it to the deck below. He had explained his fuel situation and requested the crew move the helicopters that crowded the deck so that he could land on the 'runway' (deck). MIDWAY's crew quickly cleared the deck of as many of the evacuation helicopters as possible. The O-1A, with its fuel nearly exhausted, made a perfect carrier approach, bounced twice and finally came to a safe stop; the first O-1 to ever land on a carrier deck. Upon leaving the aircraft, MAJ Boung and his family asked for and were granted political asylum in the United States. The O-1 was returned to the US and given to the Naval Aviation Museum in Pensacola, Florida, where it was displayed for a time, then refurbished and placed in storage at the museum.

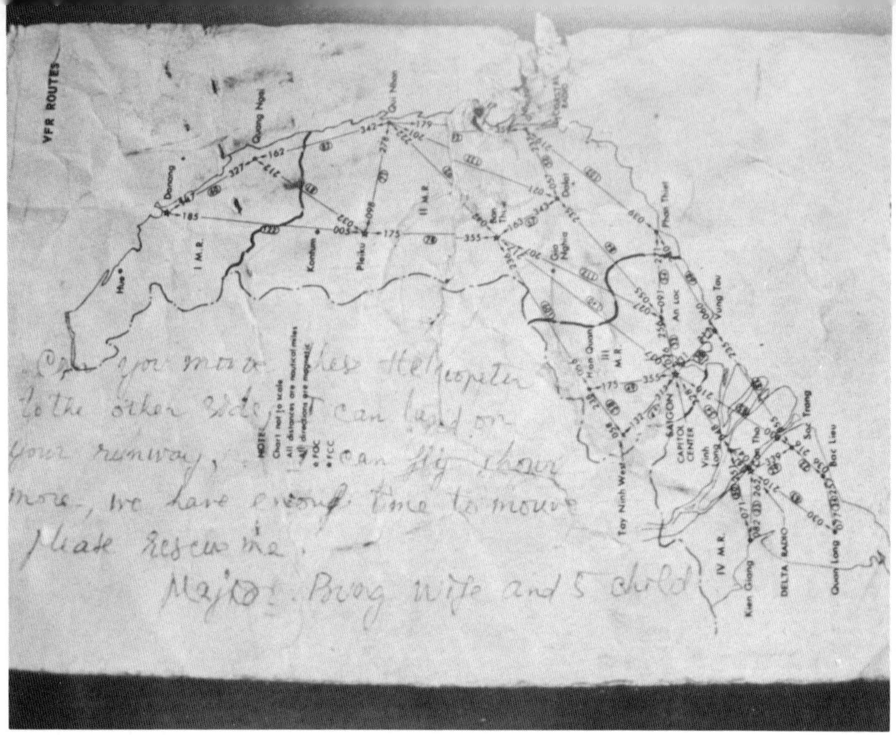

This VFR route chart was used by Vietnamese MAJ Buong to relay his plan to the crew of the USS MIDWAY. The MAJ wrote, "Can you move these helicopters to the other side. I can land on your runway, I can fly 1 hour more, we have enough time to move, please rescue me." (US Navy via R. Pickett)

With flaps fully lowered, MAJ Boung approaches the deck of the MIDWAY. The helicopters, UH-1 Hueys and AH-1 Cobra gunships, were moved to the starboard side of the flight deck to clear as much area as possible for the landing attempt. (US Navy via R. Pickett)

Landing tail wheel first, the O-1A (5-14981) bounced twice before rolling to a stop. When the aircraft halted the MAJ was greeted by a Marine guard who ensured that the aircraft carried nothing but people. (US Navy)

After a security check, the deck crew directed MAJ Boung to a parking spot. This was the first known landing by an O-1 on a carrier deck, however, before the day was out two other O-1s would find safe haven on Navy carriers. (US Navy via R. Pickett)

Surrounded by clapping MIDWAY deck crewmen and armed Marines, the MAJ, his wife, and five children are led from the flight deck for medical checks and processing. The MAJ requested and was granted relocation to the US. (US Navy via R. Pickett)

The aircraft was returned to the United States where it was displayed at the Naval Aviation Museum, Pensacola, Florida, where the Vietnamese national insignia was overpainted in Black. The leading edges of the wings and struts showed a great amount of wear and weathering. (US Navy via R. Pickett)

The O-1 was later refurbished by the Navy for display at Pensacola. The fin flash, however, was incorrectly repainted Yellow with Black stripes instead of the proper markings of Yellow and Red stripes. (Naval Aviation Museum via W. Johnson)

Foreign Bird Dogs

During its long career the O-1 Bird Dog has served the armed forces in a number of foreign countries. Chile was the first, purchasing, through the US Military Assistance Command, six L-19As. Canada soon followed ordering sixteen L-19As for the Canadian Army. These Bird Dogs, like the Chilean order, were purchased through the USAF.

France took delivery of ninety L-19As during the 1950s for service with the French Army. Many of these aircraft saw duty in Indochina and after the French withdrawal, were passed to the Republic of South Vietnam Air Force (VNAF) and the Royal Laos Air Force during the mid-1950s. Cambodia received a number of L-19s from the US for service with the Royal Khmer Aviation in the liaison role.

American assistance to the Vietnamese Air Force (VNAF) was substantial and during the mid-1950s sixty L-19s were supplied to the VNAF under the Military Assistance Program (MAP). As the war escalated additional L-19s were made available, both to increase the strength of the VNAF and to make up losses. Eventually nine VNAF liaison squadrons were flying L-19/O-1s. The majority of the L-19/O-1s provided to Vietnam were either destroyed or taken over by the Socialist Republic of Vietnam Air Force (SRVAF) after the fall of South Vietnam on 30 April 1975. There have been reports that these aircraft have been used by the SRVAF for operations over Cambodia, working with Mi-24 Hind helicopters in search and destroy missions against Cambodian resistance fighters.

The Japanese were supplied a number of L-19s by the USAF after the Korean War ended in 1953. The Japanese found the L-19 well suited to their needs and Fuji Heavy Industries of Tokyo contracted with Cessna to produce the L-19 in Japan. The L-19s built in Japan were supplied as 'knock down' components kits from Cessna, requiring only final assembly.

The Republic of Korea (ROK) was supplied with a number of L-19s after the Korean War as part of the Military Assistance Program (MAP) to Korea. The ROK Army later deployed a number of these aircraft to Vietnam with the 'Tiger' Division which fought alongside US troops in Vietnam. During the Vietnam War period Thailand was also supplied with a number of L-19s from USAF stocks. These aircraft saw action in border skirmishes with Cambodia and Laos.

The FACH used two O-1As as glider tugs painted in a high visibility scheme of Red and White at Los Condes Air Base. The additional air intake scoop above the propeller hub was used to provide additional cooling air for operations out of high altitude air fields. (D. Moraga)

Chile was provided with six O-1As under the Military Assistance Program during 1963. The Chilean Air Force (FACH) operated this camouflaged O-1, FACH 322, in the liaison role along the border with Bolivia, Peru, and Argentina. (D. Moraga)

Pakistan, through the MAP, was supplied with sixty L-19E/O-1Es for service with the Pakistani Army. Although their numbers have been reduced through attrition, it is believed that there are some forty O-1s remaining in Pakistan.

Italy received a number of L-19s under MAP during the late 1950s. The Italians, like the Japanese, saw potential in the L-19 and SIAI-Marchetti of Rome undertook to produce a turbo-prop variant. Before production ended in 1972, some 100 SM-1019s were built for use by the Italian Army. Austria was another European nation supplied L-19s under MAP for use by the Austrian Army. The majority of these remain operational and are based at Linz-Horsching, Austria.

Bird Dogs were also supplied to a number of other nations under various Military Assistance programs including; Brazil, Indonesia, Kenya, Lebanon, Norway, Spain, Turkey and The Philippines.

This highly polished natural metal Canadian Army L-19E (O-1E ex-USAF serial 56-4035) was one of nine provided to Canada during the late 1950s under the Military Assistance Program. Before delivery the aircraft were completely refurbished by Cessna. (D. Menard)

Canadian O-1s were equipped with skis and used for both liaison and training. This O-1A (16703, ex-USAF 53-8052) is overall Flat Olive Drab with a Black anti-glare panel, while the prop, underwing pylons, and skis are painted in Aluminum. (Canadian Forces via B. Best)

This O-1E (Cessna number 24734) coded 'CMM' was used by the French Army in the liaison role. The French purchased a total of 126 Bird Dogs, 90 O-1As and 36 O-1Es. Recently the French declared the O-1s surplus and a number of ex-French aircraft are being returned to the United States. (G. Kamp via R. Pickett)

Overall Dark Olive Drab VNAF O-1As line the runway at Bien Hoa during 1966. There was no standard size for the VNAF markings and the national insignia, aircraft codes, and fin flash were applied in the style decided by the area commander or province chief. (G. Adcock)

Coded 'D P', this VNAF O-1G (51-12236) is overall Light Gray with a Black anti-glare panel, Yellow and Black checkered fuselage band, and Medium Blue serial numbers and codes. Reportedly, VNAF O-1 pilots were under qualified and lacked the authority to call in strikes without higher approval. (USAF via D. Menard)

This unmarked Bird Dog (1674) was one of a number of L-19A/O-1As supplied to the Royal Lao Army during the late 1960s. These aircraft were upgraded with the latest radio equipment and operated out of Udorn Air Base, Thailand. (R. Pickett)

This overall Olive Drab Royal Thai Army O-1E (56-2590) was one of a number of O-1s provided under MAP during the early 1960s. The White fuselage code 'NU' indicated an Army aircraft and was repeated on the wings. The Thais absorbed a number of ex-VNAF O-1s and are now the largest operator of O-1s in the free world. (G. Kamp via R. Pickett)

Fuji Heavy Industries of Tokyo assembled L-19Es (O-1Es) for the Japanese Ground Self Defense Force using 'knock-down' kits provided by Cessna. This overall Gloss Olive Drab O-1E (11202) of the 8th Squadron based at Tsuiki, Japan during November of 1980 had the tip of the rudder in Yellow-Orange while the tips of the wings and horizontal stabilizers were in Day-glow Orange. (G. Kamp via R. Pickett)

The Italian Army purchased a number of O-1Es through the Military Assistance Program during 1962. This Italian Army O-1E (61-2985) about to land at its home base of Viterbo, Italy is camouflaged in Dark Earth and Dark Green. (Claudio Maranta via R. Pickett)

The Bird Dog Today

Although withdrawn from active military service the Bird Dog continues to play a vital role in American aviation with the Civil Air Patrol (CAP). Additionally, the International Bird Dog Association (IBDA), provides information to Bird Dog owners and Ector Aircraft provides rebuilt Bird Dogs to general aviation pilots for sport and commercial use.

The Civil Air Patrol began receiving military surplus Bird Dogs during the late 1960s in the same manner as the L-4 and L-5 were supplied to the CAP after WWII and Korea. The Civil Air Patrol was formed during early World War II as an auxiliary organization to take some of the workload off of the Army Air Corps and Navy. These civilian pilots were utilized to perform duties such as search and rescue, coastal observation, and anti-submarine search. Originally CAP pilots flew their own aircraft (without pay), however, in the Summer of 1942, they began flying Army and Navy aircraft in the anti-submarine role. The CAP proved very effective in this role, bombing some sixty submarines and receiving credit for two sinkings in the Atlantic. The CAP was put under the operational control of the US Army Air Force during 1943, operating under the USAAF for the duration of the war. Today the CAP serves as a USAF auxiliary, primarily in the Search and Rescue (SAR) role. CAP pilots have found the Bird Dog ideally suited to the SAR role and a number of downed airmen owe their lives to a CAP pilot flying an aging O-1 low and slow on a search mission. These pilots are all volunteers, serving without pay, doing an invaluable service for all who fly.

The International Bird Dog Association (IBDA) headquartered in Albuquerque, New Mexico, was formed by a group of individuals dedicated to preservation, restoration, and operation of Bird Dogs. They offer a newsletter, aircraft locator service, and conduct Bird Dog 'fly-ins' throughout the United States.

The Civil Air Patrol was supplied with hundreds of surplus O-1As after the Vietnam War, such as this O-1A (N5340G) of the Illinois wing of the CAP. The CAP was formed during the early years of WWII and continues to perform the vital search and rescue role as an auxiliary to the Air Force. (Chris Adcock)

This O-1A was purchased by Earl Caudell of Gulf Breeze, Florida for use in his aerial photography business. The aircraft starred in the movie "Day of the Fall of Saigon" where it portrayed the aircraft used by MAJ Buong, his wife and five children in their escape. (Author)

Ector Aircraft of Odessa, Texas, was formed to supply O-1s for civil owners. The Ector Company completely rebuilds the airframe incorporating complete corrosion protection so that the aircraft can be operated from either skis or floats. All military equipment is removed, civil radio and avionics gear is installed, and the rear seat is made removable so that cargo can be carried. Two basic models are offered, the Mountaineer and the Super Mountaineer. The Mountaineer can be powered by one of two different powerplants, either the Continental 0-470-11 or Continental 0-470-13. The Super Mountaineer is powered by a 250 hp Avco Lycoming 0-540-A4B5 engine driving either a fixed or constant speed propeller. The Ector Company had completed a total of forty-one Mountaineer aircraft by January of 1986, however, production of the Super Mountaineer has been suspended.

These three organizations are dedicated to the preservation of the Bird Dog, and since the Cessna Aircraft Company has produced a strong and long lasting airframe, it is likely that Bird Dogs will be flying into the 21st century.

An Ector Aircraft Super Mountaineer of the Buckeye Cellulose Corporation of Perry, Florida. Buckeye uses their Green and White rebuilt O-1s to survey the thousands of acres of pines in northern Florida, looking for fires and diseased trees. (Bob Woodford)

In Country

More aircraft of the Vietnam War from squadron/signal publications

1070 F-8 Crusader in action

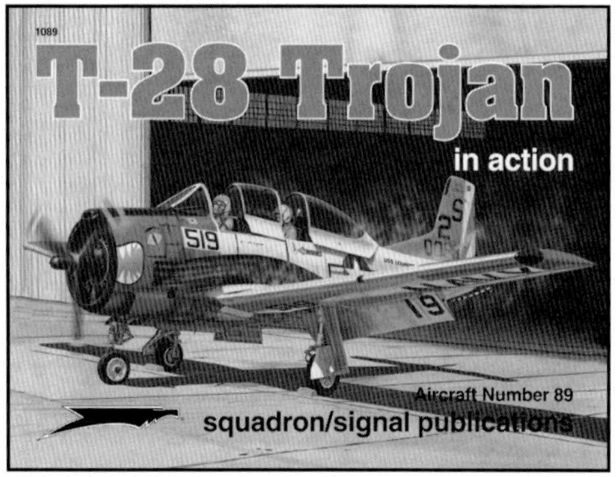

1089 T-28 Trojan in action

1120 A-7 Corsair II in action

1132 C-7 Caribou in action

1146 H-34 Choctaw in action

1161 HU-16 Albatross in action

1185 F-105 Thunderchief in action

1190 F-100 Super Sabre in action

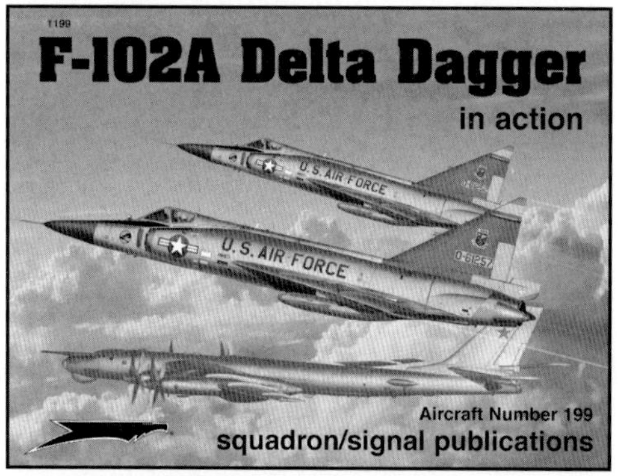

1199 F-102A Delta Dagger in action

For more information on squadron/signal books, visit www.squadron.com